NASA SP-4314

ATMOSPHERE OF FREEDOM

*Sixty Years at the
NASA Ames Research Center*

Glenn E. Bugos

The NASA History Series

National Aeronautics and
Space Administration

NASA History Office
Washington, D.C. 2000

NASA maintains an internal history program for two principal reasons: (1) Sponsorship of research in NASA-related history is one way in which NASA responds to the provision of the National Aeronautics and Space Act of 1958 that requires NASA to "provide for the widest practicable and appropriate dissemination of information concerning its activities and the results thereof." (2) Thoughtful study of NASA history can help agency managers accomplish the missions assigned to the agency. Understanding NASA's past aids in understanding its present situation and illuminates possible future directions. The opinions and conclusions set forth in this book are those of the author; no official of the agency necessarily endorses those opinions or conclusions.

Library of Congress Cataloging-in-Publication Data

Bugos, Glenn E., 1961–
    Atmosphere of freedom: sixty years at the NASA Ames Research Center/Glenn E. Bugos.
       p. cm. — (NASA history series) (NASA SP ; 4314)
    Includes bibliographical references and index.
    ISBN 0-9645537-2-4 (hardcover). — ISBN 0-9645537-3-2 (softcover).
    1. Ames Research Center—History. 2. National Aeronautics and Space
    Administration—United States.
      I. Title. II. Series. III. Series: NASA SP ; 4314.

TL862.A4 B84 2000
629.4'072079473--dc21
                                                                                               99-055757
                                                                                                    CIP

Dedicated to the people who are Ames.

TABLE OF CONTENTS

Atmosphere of Freedom
Sixty Years at the NASA Ames Research Center

Foreword
    By Henry McDonald   *vii*

Introduction   *1*

Chapter 1
    A Culture of Research Excellence: Ames in the NACA   *5*

Chapter 2
    The Transition into NASA: From a Laboratory to a Research Center   *51*

Chapter 3
    Diverse Challenges Explored with Unified Spirit: Ames in the 1970s and 1980s   *99*

Chapter 4
    A Center Reborn: Ames in the 1990s   *211*

Appendix
    Joseph Sweetman Ames   *261*

Acknowledgments   *269*

Bibliographical Essay   *271*

Endnotes   *273*

Photo Index   *277*

Index   *283*

Foreword

My hope is that you have learned or are learning a love of freedom of thought and are convinced that life is worth while only in such an atmosphere.

— Joseph Sweetman Ames,
to the graduates of Johns Hopkins University,
11 June 1935.

Ames people have won marvelous insights into the nature of atmospheres. They learned how wartime aircraft could slip through our atmosphere more precisely; how capsules could slip back into Earth's atmosphere without burning up; how airliners could wend their way safely through the congested atmosphere around airports; how to contain and control various atmospheres in wind tunnels; how the primordial atmosphere shocked into existence life on Earth; whether non-earthly atmospheres could do the same; how that earthly life has changed its atmosphere; and how to send probes to measure the atmospheres of far planets.

There's an atmosphere of freedom about Ames. There's a complex and constant convergence and intermingling of people, tools and ideas. People here approach their work with a spirit of integrity, responsibility and adventure. They value the perpetual reinvention of careers, and the cross-fertilization of ideas to solve whatever issues society faces. And they place life—from a single human operator to all the creatures in our ecosphere—at the heart of their work. Like the fog off San Francisco Bay that sometimes enshrouds the Center, the atmosphere at Ames always feels fresh, fertile, fun, and free.

Henry McDonald

Introduction

Throughout Ames history, four themes prevail: a commitment to hiring the best people; cutting-edge research tools; project management that gets things done faster, better and cheaper; and outstanding research efforts that serve the scientific professions and the nation.

More than any other NASA Center, Ames remains shaped by its origins in the NACA (National Advisory Committee for Aeronautics). Not that its missions remain the same. Sure, Ames still houses the world's greatest collection of wind tunnels and simulation facilities, its aerodynamicists remain among the best in the world, and pilots and engineers still come for advice on how to build better aircraft. But that is increasingly part of Ames' past.

Ames people have embraced two other missions for its future. First, intelligent systems and information science will help NASA use new tools in supercomputing, networking, telepresence and robotics. Second, astrobiology will explore the prospects for life on Earth and beyond. Both new missions leverage Ames' long-standing expertise in computation and in the life sciences, as well as its relations with the computing and biotechnology firms working in the Silicon Valley community that has sprung up around the Center.

Rather than the NACA missions, it is the NACA culture that still permeates Ames. The Ames way of research management privileges the scientists and engineers working in the laboratories. They work in an atmosphere of freedom, laced with the expectation of integrity and responsibility. Ames researchers are free to define their research goals and define how they contribute to the national good. They are expected to keep their fingers on the pulse of their disciplines, to be ambitious yet frugal in organizing their efforts, and to always test their theories in the laboratory or in the field. Ames' leadership ranks, traditionally, are cultivated within this scientific community. Rather than manage and supervise these researchers, Ames leadership merely guides them, represents them to NASA headquarters and the world outside, then steps out of the way before they get run over.

After twenty years as a NACA facility, Ames moved slowly into the NASA way of doing things. The life sciences came to Ames, as did new simulation facilities and heat-transfer tunnels. Yet Smith DeFrance remained as director, as distant from Washington as

Atmosphere of Freedom   Sixty Years at the NASA Ames Research Center

ever. Harvey Allen, the embodiment of the Ames spirit of scientific ingenuity, took over as director and stayed until Apollo's end was in sight. Hans Mark arrived in 1969 as a technical leader but also as an outsider. During his seven years at Ames he put an indelible stamp on the Center, retaining its scientific spirit and encouraging the tendencies toward collaboration outside the agency. In doing so, he refocused Ames' vision of itself toward broader national goals in the post-Apollo period. Then Ames stayed largely the same, while NASA gradually changed. Headquarters began to appreciate the Ames way of research management: doing projects that are faster, better, and cheaper; letting researchers freely hone the building blocks of what might someday be much larger projects; seeking

collaboration from other research institutions; and reaching out into much broader communities to bring in a diverse group of the best people. Each subsequent Center director refined and expanded that Ames culture into new areas of science and technology.

Simulation, approximation, visualization: these grander abstractions have motivated the intellectual impulses of most everyone who has worked at Ames. Ames people have simulated virtually every facet of air and space travel. Ames people built ingenious instruments to measure and model things that are not easily witnessed by the human eye: airflows, heat transfer, and the chemical compositions of far planets. They created, then overlaid, multiple methods to approximate ever better how planned devices would encounter the real world. The design of the tilt rotor aircraft—as well as of planetary probes, guided missiles, and space capsules—succeeded from constant iteration: wind tunnel tests with ever better Reynolds numbers were matched with computational fluid dynamic models having added dimensions of flows, which were

matched to controlled pilot simulations, then tested in flights approaching operational conditions. Likewise, Ames' understanding of how microgravity affects life grew through complementary terrestrial tests on animals and plants, computer modelling and controlled spaceflight experiment packages.

Ames has won many "firsts" in its scientific endeavors: thermal deicing, the blunt body concept, the supersonic area rule, hypersonic ranges, arc jets, the chemical origins of life, tilt rotor aircraft, computational fluid dynamics, massively parallel computing, air traffic controls, astrobiology, telepresence, airborne science, infrared astronomy, exploration of the outer planets, and the discovery of water on the Moon. Rather than establishing when Ames was first among its research peers, this book instead focuses on how these accomplishments contributed to the greater scientific endeavors and how they affirm and

exemplify an enduring research culture. Ames has played a pioneering role in science and technology over six decades, and all those who labored here can take pride in how they have worked together to create the atmosphere of freedom that makes excellence flourish at the NASA Ames Research Center.

A-26B bomber in the 40 by 80 foot wind tunnel.

1939 1958

A Culture of Research Excellence

Chapter 1:
Ames in the NACA

"NACA's second laboratory:" until the early 1950s, that was how most people in the aircraft industry knew the Ames Aeronautical Laboratory. The NACA built Ames because there was no room left to expand its first laboratory, the Langley Aeronautical Laboratory near Norfolk, Virginia. Most of Ames' founding staff, and their research projects, transferred from Langley. Before the nascent Ames staff had time to fashion their own research agenda and vision, they were put to work solving operational problems of aircraft in World War II. Thus, only after the war ended—freeing up the time and imagination of Ames people—did Ames as a institution forge its unique scientific culture.

With a flurry of work in the postwar years, Ames researchers broke new ground in all flight regimes—the subsonic, transonic, supersonic, and hypersonic. Their tools were an increasingly sophisticated collection of wind tunnels, research aircraft and methods of theoretical calculations.

Computers running test data from the 16 foot wind tunnel.

Their prodigious output was expressed in a variety of forms—as data tabulations, design rules of thumb, specific fixes, blueprints for research facilities, and theories about the behavior of air. Their leaders were a diverse set of scientists with individual leadership styles, all of whom respected the integrity and quiet dignity of Smith DeFrance, who directed Ames from its founding through 1965.

This culture is best described as Ames' NACA culture, and it endures today. The NACA was founded in 1915, when Americans discovered that their aircraft were inferior to those of the Europeans. The NACA itself had a unique management structure—built around a nested hierarchy of committees that served as a clearinghouse for information about the state of the art in aircraft technology. The heart of the NACA was its executive committee, supported by a main committee of fifteen, and a wide array of subcommittees formed to address specific problems. Committee seats were coveted by leaders of the aircraft industry, airlines, universities and military services. In 1917, the NACA built a research laboratory at Langley Field near Norfolk, which developed "tunnel vision" around its focus on applied aerodynamics. Whenever the NACA subcommittees could not think of a solution to some aircraft problem, they tasked the research staff at Langley to work on it. Because the NACA committees were strong, its headquarters was weak. Because the NACA was a tiny organization that carefully served the vital needs of more

USS Macon *in 1933, tethered to its mooring post after emerging from Hangar One prior to a flight from Moffett NAS*.

powerful agencies, it was largely free of political mingling.

Because of the way DeFrance patrolled the borders of his laboratory, many scientists at Ames knew little about the larger NACA context in which they pursued their work. Yet the NACA committee culture had a clear impact on the Ames research culture—the profusion of outside collaboration, belief in the value of sophisticated research facilities, appreciation of those who do good science in the cheapest and fastest way, hiring the best people and encouraging them to reinvent themselves as new research areas arose.

FUNDING THE WEST COAST LABORATORY

World War II began, for the NACA, early in 1936 when the main committee confirmed the enormity of Nazi Germany's investment in aeronautical research. The NACA learned quickly that Germany had built a research infrastructure six times bigger than the NACA's, that German universities were producing many more trained engineers, and that German aircraft might soon be the best in the world. Well before Germany invaded Poland, in 1939, the NACA was on a self-imposed war footing. Yet until then, Congress and the Bureau of the Budget kept NACA planning entrapped in Depression-era politics. The Special Committee on Relations of NACA to National Defense in Time of War, though formed in October 1936, was unable to formulate any feasible proposals until August 1938.

The Langley laboratory was simply overbuilt. Major General Oscar Westover, chief of the Army Air Corps and chairman of the NACA special subcommittee, wrote that aeronautical research was hampered by "the congested bottleneck of Langley Field."[1] Plans for upgrading the infrastructure of the base went unfunded during the early Depression, and a 1936 deficiency appropriation for new facilities quickly showed how little capacity remained at Langley. There was little room left for new wind tunnels and, more importantly, little extra capacity in the electrical grid to power them. The skies over Norfolk were

A 1938 aerial view of Moffett Field, just before construction of the NACA Ames Aeronautical Laboratory.

filled with aircraft from all the military services, and the tarmac at Langley had little extra space for research aircraft.

In October 1938, the NACA formed a new Committee on Future Research Facilities, chaired by Rear Admiral Arthur Cook. By 30 December, when Cook's committee submitted its report, the world had become a very different place. Gone was the optimism surrounding the Munich conference in September, as the Allies sacrificed Czechoslovakia in a futile attempt at appeasement. Hitler admitted that he had built an air force in direct defiance of the Versailles Treaty, and then occupied Austria without resistance in large part because of his air power. The NACA expansion plans finally rode the coattails of general military preparedness funding.

The NACA plans included some expansion at Langley, plus one new aeronautical research laboratory and a second laboratory specializing in propulsion. The NACA site selection committee had sketched out the general conditions for siting a second aeronautical laboratory: that it be on an existing Army or Navy flying field; that it offer year-round flying conditions; that it have adequate electrical power; that it be near sea level; and that it be near an industrial center for easy access to skilled labor and technical supplies. Initially the NACA preferred a location that

Atmosphere of Freedom

Admiral William A. Moffett, the architect of naval aviation, for whom Moffett Field was named.

was inland—isolated from German or Japanese attack—but then decided those fears were overcome by the need to locate closer to the West Coast aircraft industry.

They selected Moffett Field, in Sunnyvale, California. Moffett Field had been opened by the U.S. Navy in 1933 as a West Coast base for its dirigibles. The Army Air Corps took over Moffett Field in 1935, following the crash of the Navy dirigible U.S.S. Macon, and built a big airfield on the flat marsh lands in the southern portion of the San Francisco Bay. Almost half of U.S. aircraft manufacturing was located on the West Coast, within a day's rail journey from Sunnyvale. Yet it was far enough away that industrial engineers could not pester NACA researchers.

Service to industry became an ever larger part of the NACA agenda. Military procurement officers increasingly asked NACA researchers what was possible in the state of the art of aircraft design, then drew up specifications to match the NACA comments. Industrial engineers, with the task of building to these specifications, then brought to the NACA problems for solution and prototypes for testing. Since trips between southern California and the Langley laboratory consumed time and money, manufacturers turned instead to local resources, like the GALCIT wind tunnel in Pasadena.

Moffett Naval Air Station before the arrival of NACA. Dirigibles attached to a mooring post in the large circle before being wheeled into the hangar.

Therein lay the first attack on NACA plans for a second laboratory.

Since its founding in 1927, the Guggenheim Aeronautical Laboratory of the California Institute of Technology (GALCIT) had grown apace with the southern California aircraft industry. Clark Millikan, director of GALCIT, in conjunction with famed Caltech aerodynamicist Theodore von Karman, in December 1938 proposed an upgrade to their tunnel. Sensitive to the NACA territory on the spectrum of aeronautical research, Robert A. Millikan, chair of Caltech's executive committee, said this tunnel would be only for applied research, meaning for the application of theory to specific industrial designs. Millikan proposed construction of a variable density tunnel, with a 12 foot cross section, and capable of speeds up to 400 miles per hour. It would cost only $785,000, far less than the complete NACA second site.

Millikan passed his request along to General Henry "Hap" Arnold, new chief of the Army Air Corps and thus a new member of the NACA. Would Arnold fund the new tunnel at GALCIT to complement work done at Langley and Wright Field? Arnold heard his NACA colleagues argue that talk like this could derail its proposal for a second laboratory, which was working its way through the executive branch and

The first test aircraft to arrive at Ames, on 14 October 1940, was this North American O-47A.

A Culture of Research Excellence: 1939 – 1958

Atmosphere of Freedom | Sixty Years at the NASA Ames Research Center

Charles Lindbergh (left) meets with Smitty DeFrance and Jack Parsons (standing).

Congress. On the other hand, the military seemed favorably disposed to the GALCIT proposal, and the industry on the West Coast was flexing some lobbying effort in support of it.

NACA opposition to the GALCIT proposal might seem to be mere obstructionism. In postulating the research spectrum in aeronautical science over the years, the NACA had carefully divided the labor with its clients—the military services and industry—rather than contesting roles in basic science with the universities. Before the 1940s, American universities had contributed little besides broadly trained engineers to American aeronautical development. Now, Millikan again raised the relationship between academia and the NACA in a dangerous way. First, he associated the NACA with the universities on the basic side of the spectrum, separating it more clearly from the applied research it did for its clients. Second, Millikan proposed that Caltech specifically served the West Coast aircraft industry. To place government-funded research tools in von Karman's hands, NACA officials realized, was to arm a rival in a field that NACA meant to command. So Arnold sided with the NACA, decided to build a new military tunnel at Wright Field, and stopped supporting the GALCIT proposal. When the Millikan proposal failed to win Army support, Congressman Carl Hinshaw, whose district included Caltech, introduced a bill to fund a Caltech wind tunnel. Jerome Hunsaker, then chairman of the NACA, testified that Caltech appealed for government funds only because southern California firms were unwilling to fund a tunnel that would directly serve them. The proposal failed, leaving NACA even more determined to get funding for its Sunnyvale laboratory.

During World War II the U.S. Navy also built new facilities at Moffett Field, including two huge blimp hangars on the eastern side of the tarmac.

Groundbreaking for the NACA construction shack at Moffett Field, 20 December 1939, supervised by Russell Robinson (far right).

The NACA proposal cleared the next big hurdle—the Bureau of the Budget—and was forwarded to Congress by President Roosevelt on 3 February 1939. Then came the unexpected. The usually friendly House Appropriations Committee approved the expansion at Langley, but reported adversely on the Sunnyvale laboratory. This was the first congressional rejection of any major NACA proposal.

For the first time in its history, the NACA stood between a rock and the pork barrel. The long-time chairman of the House Appropriations Committee, Clifton Woodrum of Virginia, always passed along the NACA requests when they emanated from headquarters in Washington or the laboratory in Langley. The NACA never abused Woodrum's assistance, and submitted realistic estimates that were efficiently executed. Woodrum suspected, rightly, that a new laboratory in Sunnyvale would divert funding from Langley. And there were no Congressmen from California on the committee to barter pork. The NACA was unprepared to do politics this new way but learned quickly. On the day Woodrum's committee turned down the Sunnyvale request, NACA executive secretary John Victory wired to Smith DeFrance, then a Langley staffer doing advance work in California: "Entire project disapproved…. You proceed quietly and alone and learn what you can for we still have hope."[2]

Smith J. DeFrance, founding director.

The NACA started by collecting endorsements. The day after the committee's rejection, General Arnold and Admiral Cook signed a joint statement declaring that "the Sunnyvale research project is emergency in character and of vital importance to the success of our whole program for strengthening the air defense of the United States." NACA chairman Joseph Ames sent this statement to the president and tried, unsuccessfully, to have the Senate reintroduce the NACA proposal.

A Culture of Research Excellence: 1939 – 1958

Atmosphere of Freedom Sixty Years at the NASA Ames Research Center

Ames' Flight Research Laboratory, in July 1940.

So the NACA executive committee met in June and appointed a special survey committee on aeronautical research facilities, chaired by Charles Lindbergh and composed of General Arnold, Admiral John Towers, and Robert H. Hinkley, chairman of the Civil Aeronautics Authority. During the congressional rehearing of the Sunnyvale proposal, they reached a neat compromise, facilitated by the prestige of Lindbergh and the power of the other members of this special committee. Congress approved the NACA proposal for a second laboratory, but deleted the provision establishing it in Sunnyvale. Instead, the NACA had to select a site within thirty days after the bill was passed. The bill passed on 3 August, and Lindbergh's committee reevaluated its list of 54 newly proposed sites. On 19 October 1939 the Lindbergh committee settled, not surprisingly, on the Sunnyvale site. (Lindbergh's evaluation of these sites proved very useful in the fall of 1940, when his committee was also asked to select a site for a new engine research laboratory, which they located in Cleveland.)

The turmoil over establishing the NACA's second laboratory had a lasting impact on Ames. First, everyone within the NACA became even

The first two Ames staff to arrive for work on 29 January 1940. John F. Parsons (left) built a sterling reputation for constructing wind tunnels, and Ferril R. Nickle (right) kept Ames' budget and procurement practices lean and efficient.

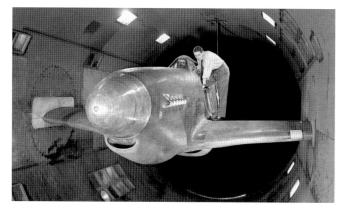

North American XP-51B airplane in the 16 foot wind tunnel, in March 1943, with the outer wing sections removed, readied for full scale studies of duct rumble.

more sensitive to the verbiage of basic and applied research, so that even today people at Ames wax fluent on their place within the research spectrum. Second, Ames staff had no time to get grounded in the place before being swept up into war work.

WAR WORK

Even before Congress had finalized its funding, the NACA was ready to start work on the Sunnyvale site. By 6 December 1939 the NACA had worked out an agreement with the War Department over 43 acres at Moffett Field tentatively called the Aeronautical Research Laboratory, Moffett Field. Ground was broken on 20 December 1939 for a solitary wooden construction shack to house the small staff on-site, supervised by Russell Robinson. Meanwhile, DeFrance returned to Langley where he was hand-picking his research staff and overseeing their designs for facilities at the new laboratory.

The 16 foot high speed wind tunnel under construction in October 1940.

The first permanent staff arrived at Ames on 29 January 1940, led by John Parsons and Ferril R. Nickle. Good memories of Stanford University convinced many Langley staffers to relocate to the new laboratory. Parsons had worked closely with William Durand, professor of aeronautics at Stanford and a leading member of the NACA. More than twenty Stanford graduates filled out the Ames staff within its early years, including Harvey Allen, Walter Vincenti, and John Dusterberry.

A Culture of Research Excellence: 1939 – 1958

Atmosphere of Freedom   **Sixty Years at NASA Ames Research Center**

In February 1940, construction began on the flight research building; in April, work started on the first of two technical service workshops; in May, work began on the 16 foot high speed wind tunnel, as well as on the first of two 7 by 10 foot tunnels. In July 1940, DeFrance took over officially as engineer-in-charge and the first test piles were dug for the 40 by 80 foot wind tunnel, larger by a third than the biggest at Langley. Research first began at Ames in October 1940, wind tunnels started returning data, and by the time of the raid on Pearl Harbor, the new laboratory had published its first technical report.

Deicing Research

The first research effort authorized at Ames focused on ways to defeat the icing menace. Icing was the major impediment to safe and regularly scheduled air transportation, and had already disrupted wartime military flights. Yet little was known about how to knock ice off an aircraft, and even less about what caused it. Lewis Rodert had already started this research at Langley, but thought the weather in northern California was better suited to the study of icing conditions. The flight operations hangar was the first research building opened at Ames, and Rodert based his research effort there. Furthermore, the NACA leaders had followed the deicing work and knew that it was close to producing

Glaze ice jutting forward on the radio antenna and airspeed pitot mast of the C-46.

Lockheed 12A icing research airplane in February 1941, with heated wings.

Lewis Rodert accepting the Collier Trophy from President Harry Truman in December 1947.

important results, which would quickly validate their fight for the new laboratory.

To really understand how ice formed on aircraft, Rodert and his group first needed to devise an aircraft that could collect data in even the worst icing conditions. As an expedient to in-flight experimentation, they tried out thermal deicing. They ran hot exhaust gas through the wings of a Lockheed 12A, and discovered that thermal deicing worked well.

After first defining the problem and refining the specific technologies of thermal deicing, Rodert rushed to devising design rules of thumb for those preparing aircraft for war. His techniques for thermal deicing were built into many aircraft important to Allied air operations in World War II, including the B-17, B-24 and various PBY naval patrol aircraft. Toward the end of the war, Rodert's group focused more on theoretical calculations of the heat required for deicing, though he continued to agitate among aircraft designers for more attention to icing problems.

Rodert won the prestigious Collier Trophy in 1947, soon after he had left Ames for the NACA's Aeronautical Propulsion Laboratory. His work was soon superseded by new technologies, especially

Ames work in thermal deicing included this laboratory test section of the electrically heated airfoil for the C-46, in November 1945. The thermocouples and nichrome electrical heating elements are already installed.

A Culture of Research Excellence: 1939 – 1958

Ames Aeronautical Laboratory in November 1940, with the flight research hangar in front and the 16 foot tunnel under construction.

Ames a year later, (facing page) with the administration building and the two 7 by 10 foot tunnels completed.

those made possible with jet engines. And his work was soon forgotten around Ames. On one hand, his deicing work was atypical of the work then dominant at Ames. Rodert had taken his work much further into practical design issues than the NACA was ever meant to go, and his work had little to do, ultimately, with wind tunnels. In fact, Rodert distrusted the ability of wind tunnels to produce artificial ice anything like natural ice. On the other hand, Rodert had started with a bold theoretical stance, and defended it tenaciously. He paid attention to his research tools, specifically the airborne laboratory that let him prove out his ideas cheaply and quickly. Thus, his research in many ways foreshadowed the Ames way of research.

Wartime Wind Tunnels

The key component in Ames' research agenda, and its first construction priority, was the 16 foot high speed wind tunnel. Opened in 1941, it proved a remarkably timely tool in

The 16 foot wind tunnel, in April 1948, viewed from the top of the 40 by 80 foot wind tunnel.

refining wartime fighter aircraft. Its test section was four times larger than Langley's 8 foot high speed tunnel and its speed, up to Mach 0.9, or 680 miles per hour, made it the fastest in the NACA and ideal for solving problems specific to air compressibility. The Lockheed P-38, for example, was the first aircraft able to fly fast enough to encounter compression effects. It had a fatal tendency to tuck under; that is, in a high-speed descent, it nosed over into a vertical dive from which no pilot had the strength to recover. Researchers at Langley investigated and found shock waves along the wing that reduced lift.

When they suggested a radical redesign of the aircraft, early in 1943, Lockheed chief Kelly Johnson instead took the problem to Ames. Led by Albert Erickson, the 16 foot group found the specific location of the shock wave and showed how it caused flow separation over the wing. This, in turn, removed the downwash on the tail to put the aircraft into a dive; no elevator had enough surface area to allow the pilot to pull out of it. While the complex of aerodynamic factors was fascinating, Ames people understood that the Lockheed engineers looking over their shoulders wanted a quick answer. Erickson explored a number of configuration changes, the simplest of which was a flap under the wing. DeFrance, in reviewing this work, suggested hinging the flap so that the pilot

The 40 by 80 foot wind tunnel under construction in June 1943. A naval patrol blimp floats in the background.

A Culture of Research Excellence: 1939 – 1958

The 40 by 80 foot wind tunnel, newly opened in October 1948. The test section is the square building in the center. Adjacent is the technical services building, the utilities building, and the 16 foot wind tunnel.

could control the dive. From these insights, Ames developed dive-recovery flaps which were immediately built into the P-38 and the Republic P-47 and later added as safety devices for flight tests of all new fighter aircraft.

Duct rumble on the P-51B Mustang—another example of the utility of the 16 foot tunnel—was so bad that, at 340 miles per hour, flow through the inlet caused the aircraft to buffet dangerously. The president of North American Aviation made an emergency appeal to DeFrance, and one week later the P-51B fuselage was mounted in the 16 foot tunnel and ready for tests. Within two weeks, Ames engineers had successfully modified the shape of the duct inlet. Engineers at North American built inlets according to Ames' design, finished the flight tests, and the P-51B went on to become the fastest and most potent fighter plane in Europe.

There was nothing especially sophisticated about Ames' twin 7 by 10 foot wind tunnels. "Workhorse" was how they were most often described. But from the time they opened in the fall of 1941 they were kept in almost constant use, mostly to correct design faults in new military aircraft like the B-32 and the XSB2D-1. Because models used in these low-speed tunnels could be made entirely of wood, it was cheap and easy to run tests there. Ames staff always found ways to squeeze time from the 7 by 10 foot tunnels for basic research. There they pioneered the use of electrical motors on models to simulate propeller flows, then studied the debilitating effects of propeller slipstreams.

Many of the 7 by 10 foot tunnel staff moved over to the 40 by 80 foot wind tunnel when it opened

Control room of the 40 by 80 foot wind tunnel soon after it opened. The scales measured lift, drag and other forces mechanically.

Ames messengers, 1945

in June 1944. Harry Goett led the new full-scale and flight research branch, which included the research aircraft. The 40 by 80 foot tunnel was best suited to aircraft development work, rather than basic research. The first series of tests was for the BTD-1 Destroyer, a rather ambitious fighter designed by Douglas Aircraft. After countless hours of testing at Ames, the Navy lost interest in the BTD-1 as the war came to a close. Other aircraft tested there included the Northrop N9M-2 flying wing prototype, the Grumman XF7F-1 Tigercat, the Douglas A-26B low-level bomber, and the Ryan XFR-1. Where the 40 by 80 foot wind tunnel distinguished itself most was in the study of complex airflows and handling qualities at slow speeds.

Flight research complemented all facets of Ames tunnel research, and Ames aerodynamicists constantly checked data generated in the wind tunnels to see how well it agreed with data generated in free flight. For example, Ames staff, working at the NACA high speed flight research center at Rogers Dry Lake, California, calibrated tunnel and flight data using their P-51 aircraft. They removed the propeller from the P-51 so the aircraft

Northrop P-61A Black Widow towing a North American P 51B from which the propeller was removed for data calibration tests.

Ames pilots in June 1942 (left to right): Larry Clousing, Bill McAvoy and Jim Nissen.

A Culture of Research Excellence: 1939 – 1958

Grumman F6F-3 Hellcat modified, in 1948, by Ames engineers to become the world's first variable stability aircraft.

would be aerodynamically clean like a tunnel model. Another aircraft towed it to altitude, released it, and Ames test pilot James Nissen guided it to a landing while recording airflow data. Drag flow and all other measurements correlated superbly with data generated in the 16 foot tunnel.

Handling Qualities

With the wealth of data collected on the P-51 flights, Ames engineers moved into research on handling qualities. During the war, Ames had tested a wide array of different military aircraft in its 7 by 10 foot tunnels. Although these tunnel tests were meant to solve specific problems of stability and control, the Ames aerodynamicists began to see patterns in the problems. Ames test pilot Lawrence Clousing, working with William Turner and William Kaufmann, led early efforts at describing in objective and universal terms the handling qualities of aircraft for handbooks on specific aircraft. In the early 1950s, Ames investigated handling qualities more systematically in order to develop a guide for evaluating new military aircraft. Three Ames pilots flew ten different aircraft in 41 different configurations to determine, first, the safe minimum approach speed for aircraft landings and, second, any more general stability and control issues. From these test flights, pilot George Cooper devised a standard ten-point scale for rating handling qualities that assessed the difficulty of maneuvers, the aircraft's behavior and pilot accuracy. The Cooper Pilot Rating Scale, published in 1957, standardized handling qualities assessments across the industry and around the world. (It was revised in 1969 by Robert Harper of Cornell Aeronautical Laboratory, and is now called the Cooper-Harper Handling Qualities Rating Scale.)

Anti-turbulence screen in the 12 foot pressurized wind tunnel.

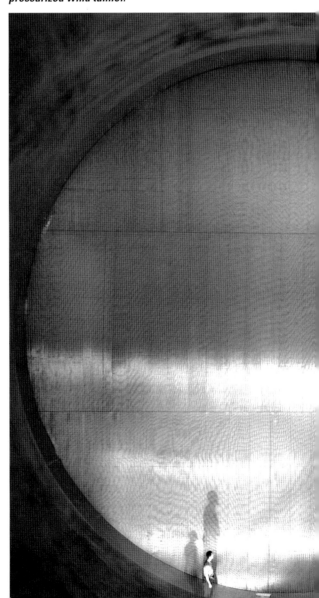

The 12 foot pressurized wind tunnel, newly completed in 1943. The turbulence screen is in the big ball in front of the cubical test section building.

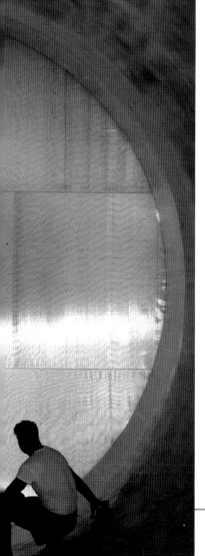

The Ames flight research group also pioneered variable stability aircraft. In 1948, a group led by William Kaufmann altered a Grumman F6F-3 fighter by adding servo-actuators to the ailerons so that the pilot could modify the dihedral of the wing (whether it slants upward or down). They added a drive to the rudder so the pilot could vary directional stability and damping, and soon devised other mechanisms so the pilot could vary six key stability and control parameters. For the first time, aerodynamicists could change flying qualities, even in flight, without changing the aircraft's configuration. Ames aerodynamicists could easily explore flying qualities of any aircraft then under design. For example, as a result of pilot comments during variable stability tests on the F6F-3, Ames suggested that Lockheed design the F-104 with ten percent negative dihedral. To improve military specifications on flying qualities, Ames later applied the concept to such aircraft as the F-86D, the F-100C, and the X-14.

12 Foot Pressurized Wind Tunnel

Ames' most sophisticated facility for calibrating tunnel tests with free flight was the 12 foot pressurized, low turbulence tunnel. It opened in July 1946, and stood as the culminating achievement of subsonic tunnels. Pressurization directly addressed the issue of Reynolds numbers. Is one justified in drawing conclusions about the properties of large bodies, like aircraft, from tests on smaller objects, like models? That is, are there scale effects because of the thickness of air? A Reynolds number is a statement about the relationship between the four properties that affect the flow of a fluid about a body moving through it—the size of the body, and the air's velocity, density and viscosity—and most simply it expresses the ratio of aerodynamic forces to inertial forces. Tunnel tests are comparable only when the Reynolds numbers are the same. To get numbers to compare tunnel scale models with full size aircraft, researchers must make the air in the tunnel more dense. Thus to compare data from an aircraft flying at 800 miles per hour, with a

Smith DeFrance greets his staff as they prepare to have their picture taken, 30 August 1940.

one-fifth scale model aircraft also at 800 miles per hour, the air pressure must be raised fivefold. This was the thinking behind the variable-density, pressurized wind tunnel.

Building the pressurized tunnel was an engineering marvel. Because the hull had to withstand five atmospheres of pressure, the steel plates in some places of the hull were two inches thick. The pylons on which the 3,000 ton hull was mounted were hinged to allow for expansion during heating and pressurization. Instead of the usual sharp 90 degree angles to turn around the airstream, the hull turned it around in small angular steps. Finally, to improve the uniformity of the flow, Ames built a 43 foot diameter sphere just before the test section to hold a fine-mesh anti-turbulence screen.

The 12 foot tunnel was used immediately to explore the performance of low aspect ratio wings, swept wings and delta wings like those used in the Air Force's Century series of fighters. And it was used in basic research where scale effects mattered—like in the design of wing flaps and laminar flow control devices. Most important, it allowed closer correlation between results from wind tunnels and flight tests.

DEFRANCE, PARSONS, GOETT AND ALLEN

The Ames work force grew rapidly during the war and afterward, from 50 in 1940, to 500 by 1943, to 1,000 by 1948. As the number and variety of researchers at Ames expanded, its organizational chart grew more complex. However, the structure of leadership at Ames remained fairly clear. During Ames' first two decades, four men formed the contours of its organizational culture—Smith J. DeFrance, John F. Parsons, Harry J. Goett, and H. Julian Allen.

John Parsons presenting to Smith DeFrance (seated) a 35 year service award, in July 1957, that had just arrived in the mail.

Full-scale Douglas XSB2D-1 airplane mounted in the 40 by 80 foot tunnel.

Smitty DeFrance, the director, was a pillar of integrity, a conscience of conservatism, and a reminder that everyone at Ames worked for a greater good. DeFrance had served as a pilot during World War I then earned a bachelors degree in aeronautical engineering from the University of Michigan. He joined the Langley Laboratory in 1923, designed its 30 by 60 foot tunnel, and rose to lead its full-scale wind tunnel branch. Before Congress had even funded Ames, he led design studies for its first tunnels, and was named the laboratory's founding engineer-in-charge. DeFrance received the Presidential Medal of Merit in 1947 for designing and building the laboratory. His title was changed to director, a position he held until his retirement in October 1965.

DeFrance stayed close to the Ames headquarters building, where few of his staff ever went. DeFrance's management style has been described as that of a benevolent dictator who patrolled Ames' boundaries. The NACA headquarters largely demanded that of its directors: only one voice should speak for the laboratory so all contact and correspondence went through his office. In turn, he shielded his research staff from outside pressures, created an atmosphere of freedom, and allowed the laboratory to evolve like a think tank. When DeFrance did have contact with his research staff, it was to inquire about contingency or emergency plans, the public value of a project, or how certain his staff was of their conclusions.

Perhaps because he had lost his left eye in an airplane crash at Langley, DeFrance insisted on extraordinary safety measures. It was DeFrance who insisted that the

Atmosphere of Freedom   Sixty Years at the NASA Ames Research Center

pressure hull of the 12 foot wind tunnel be tested hydrostatically—that is, by filling it with 20,800 tons of water to see if it would burst. Later, DeFrance was in the control booth as engineers cautiously started turning the fan blades for the first time. "What's that red lever for?" DeFrance asked above the rising roar of the motors. "An emergency shut-off," yelled back an engineer. DeFrance leaned over and pulled the lever. The engineers just stared as the fragile blades shuttered to a halt. "Don't you think you should be sure that the shut-off works," Smith said, "before you need it?"[3] No one ever questioned DeFrance's experience.

Also because of his airplane accident, DeFrance promised his wife he would never fly again. Since the train trip to Washington took four days each way, he

Ames people, 30 August 1940, in front of the new flight research building. First Row: Mildred Nettle, Margaret Willey, M. Helen Davies, Marie St. John, Smith DeFrance, Edward Sharp, Manie Poole, Virginia Burgess, Roselyn Pipkin. Second Row: Arthur Freeman, Thomas O'Briant, Lesslie Videll, Clyde Wilson, Mayo Foster, Manfred Massa, Manley Hood, Carlton Bioletti, Charles Frick, Walter Vincenti, Howard Hirschbaum, Lewis Rodert, Eugene Braig, Carl Gerbo. Third Row: Rowland Browning, Donald Hood, Robert Hughes, George Bulifant, James Kelly, Harvey Allen, John Houston, Karl Burchard, Mark Greene. Fourth Row: Andre Buck, Edward W. Betts, Raymond Braig, Harry Goett, John Parsons, Herbert Dunlap, Lysle Minden, Frank Clarke. Fifth Row: Walter Peterson, Wilson Walker, Charles Harvey, John Delaney Jr., Thomas Macomber, Alan Blocker, Noel Delaney, Alvin Hertzog, Ferril Nickle, Paul Prizler, Ross Benn, Edward Schnitker.

Below: Ames on 3 July 1945, toward the end of World War II: (1) administration building; (2) science laboratory; (3) technical service building; (4) 40 by 80 foot wind tunnel; (5) electrical substation; (6) 12 foot pressure wind tunnel; (7) utilities building; (8) 16 foot high speed wind tunnel; (9) 1 by 3 foot supersonic wind tunnel; (10) 7 by 10 foot wind tunnel number 1; (11) model finishing shop; (12) 7 by 10 foot wind tunnel number 2; (13) flight research laboratory; (14) airplane hangar and shop.

seldom went there. This created a curious situation in that the person responsible for speaking for Ames with NACA headquarters and other federal agencies actually did so rarely. Yet when DeFrance did speak to people in Washington, they listened. As the younger scientists at Ames grew more ambitious after the war, they often felt that their colleagues at Langley took unfair advantage of their proximity to Washington to press their own plans. In fact, DeFrance knew that distance also had its advantages in creating space for basic research. Plus, DeFrance had ambassadors in key places. In 1950, Russell Robinson returned from NACA headquarters to serve as Ames' assistant director alongside Carlton Bioletti. Robinson, especially, continued to improve Ames' relations with Washington. Edwin Hartman, who served from 1940 to 1960 as the NACA's representative among the airframe manufacturers of southern California, served as DeFrance's ambassador to the various facets of the aerospace industry.

Jack Parsons was the builder. He arrived at Ames in January 1940 with the pioneer detachment from Langley. He oversaw the entire construction effort, became DeFrance's principal assistant, and stayed as associate director until his retirement in 1967. A native of Illinois, Parsons moved to Stanford University to take a bachelor degree, as well as the professional degree of engineer, and to work with William Durand in editing his classic six-volume work titled *Aerodynamic Theory*. Joining Langley in 1932, he oversaw the design and construction of the 19 foot pressure tunnel. At Ames, in addition to serving as chief of the construction division, Parsons became chief of the full scale and flight research division.

George Cooper, test pilot, and a North American F-100 Super Sabre.

Though trained in aerodynamics, Parsons had an intuitive understanding of how to pour concrete, weld steel, and get every part of a construction team pulling together. He was the exemplar of the Ames project management style, able to complete projects on time, with ingenious engineering twists that saved money and kept the scientific results foremost. NACA headquarters turned to Parsons to lead its Unitary plan wind tunnel effort, which was conceived as the biggest single construction project in NACA history. As construction around Ames slowed its breakneck pace, Parsons turned his attention to the administration of the laboratory. As chief administrator, he saw the big picture and brooked no inefficiency. To the junior staff his presence served as a constant reminder that they, indeed, worked for the federal government, with an ultimate responsibility to the American public. He was a quiet operator, intensely loyal to DeFrance, and widely respected for his skills.

Harry Goett championed applied research and served as an early model of career reinvention at Ames. A native of New York, he earned his degree of aeronautical engineer from New York University at the nadir of the Depression. He worked at a handful of companies as a mechanical engineer before joining DeFrance's branch at Langley in July 1936. He arrived at Ames in July 1940, designed model supports, then directed research in the 7 by 10 foot workhorse tunnels. He took charge of the 40 by 80 foot wind tunnel when it opened in 1944, and in 1948 he took over Parsons' role as leader of all full scale and flight research. He remained there until July 1959, when he was named founding director of the new Goddard Space Flight Center.

Harry Goett, with Larry Clousing (seated).

Goett understood that he supervised the most sought-after set of research facilities in the world, and he strove constantly to keep them in good use. Aircraft companies might ask Goett's group to solve routine problems of control and stability, but Goett never allowed his people to see their work as routine. He constantly urged them to envision new opportunities for basic research, and to look at the bigger picture of what they were learning. This ability to see new patterns in routine work led to Ames' long-running work in handling qualities and variable stability aircraft.

Goett moved his group into research on space vehicles long before that work fell under the NACA's purview. He encouraged Jackson Stadler to pursue plans for a low density wind tunnel, opened in 1948, to explore aerodynamics where there is little air. Goett became the NACA's technical liaison to the West Coast manufacturers of satellites and space probes, and became an expert on launch systems and instrumentation for space systems. While remaining firmly within the management ranks, Goett had reinvented himself as an expert on space technology.

Goett kept his staff alert and moving ahead. He made his people understand why they were running every test, starting with a complete

The 8 by 8 inch supersonic wind tunnel, built in 1946, served as a prototype of the 6 by 6 foot wind tunnel. This detailed view of the test section shows the test mounts; the sliding block throat is set to the highest Mach number.

A Culture of Research Excellence: 1939 – 1958

A tailless delta wing aircraft, the Douglas F4D-1 Skyray, shown during flight tests in April 1956 with Ames pilot Donovan Heinle, engineer Stewart Rolls and crew chief Walter Liewar.

analysis of the problem and using the best tools of aeronautical science, so that the tunnel tests simply provided numbers for the tables. His infamous bi-weekly meetings for each branch in his division took on the air of inquisitions, as peers questioned every part of an investigation. Sharing trepidation over their day on the block built substantial esprit de corps. Often Goett suggested a novel way to resolve an intractable problem, though his name appeared on far fewer research papers than he contributed to. He was never in competition with his staff.

As a person, Goett took pride in the profession of engineering, and got along well with pilots. He was cut from the same mold as DeFrance, straight-laced, soft spoken, pragmatic, and authoritative. By contrast, there was Harvey Allen, and the men who followed his lead were a very different breed.

Harvey Allen pushed the limits, in scientific creativity as well as in social behavior. Allen was emotionally involved with his work. He never let the paperwork thrust upon him during his rise through the ranks interfere with his compulsive urge to explore the nature of air himself. This endeared him to the growing numbers of researchers at Ames.

Allen was born in Illinois, in 1910 and, like so many early Ames employees, earned his bachelor and professional engineer degrees from Stanford University. Upon graduation in 1936, he joined the NACA at Langley and developed a general theory of subsonic airfoils that

Harvey Allen, chief of Ames' High Speed Research Division in 1957.

This Lockheed YP-80A Shooting Star arrived at Ames in September 1944. As the first jet aircraft at Ames, it was used in a variety of research problems—on compressibility effects, aileron buzz, boundary layer removal and tail-pipe heating.

helped to dramatically improve low drag airfoils. Allen moved to Ames in April 1940 to lead the theoretical aerodynamics section, reporting to Donald Wood. Allen spent as much time designing as using the new wind tunnels. He conceived many of the throat designs and turbulence screens that allowed Ames wind tunnels to reach faster speeds with better results. In July 1945, Allen was named chief of Ames' new high speed research division, where he remained until further promotion in 1959.

High speed meant supersonics and hypersonics, speeds that were then only theoretical. Allen developed a now well-known theory for predicting forces at supersonic speeds at various angles of attack, a theory that proved especially useful in designing missiles. He devised theories of oscillating vortices, of heat transfer and boundary layers, and of the interaction between shock waves and boundary layers.

But Allen was no mere theoretician. He knew it would take decades for theories of supersonics and hypersonics to catch up with the reality he would forge in the meantime. Allen designed two types of

Cutaway view of a YP-80 model to be tested in the 1 by 3.5 foot supersonic wind tunnel. Building tunnel models required precise machining on the outside and compact and ingenious instrumentation on the inside.

A Culture of Research Excellence: 1939 – 1958

Harvey Allen in his Palo Alto home.

supersonic nozzles that made Ames' wind tunnels more flexible and effective. He designed two methods for visualizing airflows at supersonic speeds, and devised techniques for firing a gun-launched model upstream through a supersonic wind tunnel.

Allen will be remembered best for the insight known as the blunt body concept for solving reentry heating. He published a paper in 1951, jointly with Alfred Eggers, in which they suggested that a blunt shape was better than a pointy shape for getting a body back into Earth's atmosphere without it burning up. This insight was counterintuitive. Most other researchers assumed that a design should minimize the contact between object and air to reduce the heating; Allen and Eggers knew the air would carry away its own heat if all the shock waves were designed right. Having advanced his theory, Allen marshalled every possible resource to prove it. He built wind tunnels capable of hypervelocities, arc jets capable of high sustained heat, and flight research vehicles that pushed the envelope of space. Every human-made object that reenters Earth's atmosphere—ballistic missiles, manned space capsules, the Space Shuttle orbiter—does so safely because of Allen's passion for his research.

There were a great many giants in these formative years of Ames history—Helen Davies became division chief for personnel; Marie St. John was DeFrance's administrative assistant; Larry Clousing, Bill McAvoy, Steve Belsley, and Alun Jones ran flight operations; Donald Wood and Manley Hood ran the theoretical and applied research division; Dean Chapman and Max Heaslet were world-renowned theoreticians. To the world outside, DeFrance and Parsons were the face of Ames. But those working Ames' wind tunnels placed themselves in either Goett's or Allen's camp. And the (always friendly) tension between Goett and Allen defined the character of the place.

Where Goett had a passion for excellence, Allen had a passion for ingenuity. Said Bill Harper, who took over the 40 by 80 foot wind tunnel from Goett: "The educational impact on a young engineer, caught between these two, each arguing his case in a most convincing way, was enormous. To strengthen his case, Harvey was always holding parties at his

Lockheed NC-130B modified with boundary layer control for studies of short takeoffs and landings.

home which quickly turned into intense technical arguments....No matter who you worked for, you could expect to find Harvey dropping by to learn of your progress and constructively criticize what you were doing."[4]

Harvey Allen was a modern renaissance man: a lifelong bachelor, a world traveller, collector of ethnic arts, a lover of fine automobiles, a bon vivant with a creative and cultured mind, a hard drinker and host of legendary parties. He animated lunchtime conversations at the Ames cafeteria. Allen had a warm sense of humor that blended nicely with his highly creative mind and his informal and sincere approach to people. Allen's final research project was on the slender feather protruding in front of an owl's wings, which he suspected enabled owls to fly so silently. As a testimony to how much fun Allen made Ames as a place to work, a group of Ames alumni continue to meet, calling themselves The Owl Feather Society.

Harvey Allen had a nickname for everybody, often the same name. After he went a year of calling everybody Harvey, after the character in a popular play, the name stuck to him. (The H in H. Julian Allen was for Harold. His family called him Julian.) In 1952, Ames hired a mathematician with a Ph.D. and Allen started greeting everyone jovially as "my good doctor." One day his group sat at the start of a meeting and in

Thrust reverser on an F-94C Starfire. Discussing the flight evaluation tests, in the summer of 1958, are (left to right) Air Force Major E. Somerich, Ames engineer Seth Anderson, Lt. Col. Tavasti, and Ames chief test pilot George Cooper.

A Culture of Research Excellence: 1939 – 1958

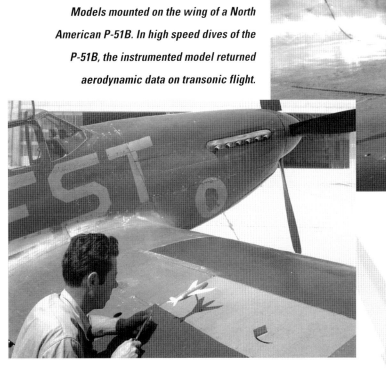

Models mounted on the wing of a North American P-51B. In high speed dives of the P-51B, the instrumented model returned aerodynamic data on transonic flight.

walked Milton Van Dyke who was young and looked even younger. When Allen called out, "My good doctor Van Dyke!" the mathematician, who had not yet caught on to Allen's conviviality, exclaimed, "My god, does everyone here have a doctorate?"[5] Those there broke into laughter. In fact, none of them had doctorates, and it didn't matter. The atmosphere was open to anyone with good ideas.

INTO SUPERSONICS

In May 1944, DeFrance and Allen first proposed to NACA headquarters a supersonic tunnel with a test section that was big enough for a person to work in. Researchers using the 1 by 3 foot supersonic tunnel could detect shock waves, but they could not use models big enough to collect pressure distributions. NACA headquarters shelved the plan for

Wing planforms of various shapes and sweep were tested, in 1948, in the 1 by 3 foot wind tunnel to determine the most efficient wing design for supersonic aircraft.

Charles Hall displaying tunnel model AR2, in February 1957, which incorporates conical camber as the half-cone twist in the wings.

the larger tunnel, claiming lack of funds. Some months later an engineer from the Navy Department showed up seeking advice on a supersonic tunnel they had hoped to build. Headquarters staff, looking prescient indeed, pulled the Ames design out of a drawer, and by January 1945 the Navy had transferred funds to get this tunnel built. Carlton Bioletti immediately started the detailed design, and the 6 by 6 foot supersonic tunnel made its first trial run on 16 June 1948. Charles Frick ran the tunnel, which was used to test every major jet aircraft and guided missile of the 1950s—for drag reduction, stability and control, and inlet design.

However, researchers were annoyed that the tunnel could not obtain data in the transonic range: it operated subsonically from Mach 0.6 to 0.9, and supersonically from Mach 1.2 to 1.9. Charles Hall led studies on a modification of the tunnel, completed in April 1955, that produced speeds continuously from Mach 0.65 to Mach 2.2. As faster tunnels, like the Unitary, came on line for development tests of operational aircraft, Ames used the 6 by 6 foot tunnel more for basic research in conical camber, vortex flows, canard-type controls, and inlet design for supersonic speeds.

Ames pioneered another facility for gaining data on transonic aerodynamics. In 1946, the Ames flight engineering section, led by Alun R. Jones, devised a way to build free-fall models and recover them, at a fraction of the time and cost of building rocket-boosted models. Ames developed 107 models and recovered 95 of them. These were mostly full-scale models weighing up to a ton. After the flight tests, the models decelerated from transonic speeds so that a parachute could deploy, then landed on a nose spike which penetrated the ground. These models proved important in validating data on transonic drag-rise which led to the theory, developed by Robert T. Jones at Ames, of

A Culture of Research Excellence: 1939 – 1958

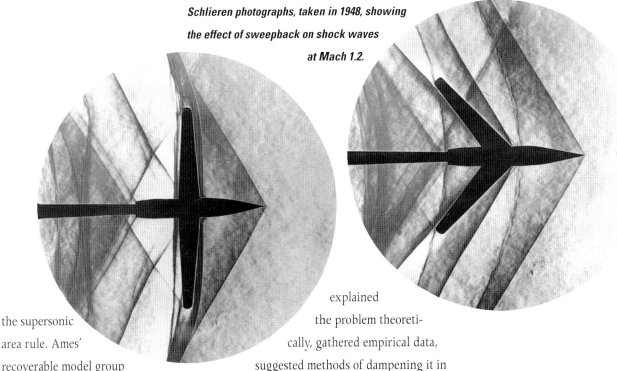

Schlieren photographs, taken in 1948, showing the effect of sweepback on shock waves at Mach 1.2.

the supersonic area rule. Ames' recoverable model group established a method for calculating optimal fuselage shapes at specified speeds and showed, by comparison, how tunnel walls and Reynolds numbers skewed design data. And they measured the values of engine air inlets, which must be tested at full size because of their extreme sensitivity to boundary layers.

Solving Jet Problems

The Lockheed P-80 Shooting Star was the first American airplane designed from scratch for jet propulsion, and thus the first to encounter the problem of transonic flutter—a fast vibration in the ailerons. Using the full speed of Ames' 16 foot tunnel, researchers first discovered that the wing did not generate this aileron buzz, as it traditionally did. Then they explained the problem theoretically, gathered empirical data, suggested methods of dampening it in other aircraft, and flight tested their ideas. Wing-body-tail interference, as another example of how Ames solved problems of supersonic flight, arose because jet bodies and tails were larger relative to the wing in order to provide stability over a wider range of speeds. Jack Nielsen led a group devising interference theories that were tested by comparing theoretical results with tunnel data.

Robert T. Jones, theoretical aerodynamicist.

A swept wing model readied for a test, in June 1948, in the 1 by 3 foot blowdown tunnel with a variable geometry throat mechanism.

Ames' work on supersonic aircraft focused first on the sort of work the NACA had always done for America's aircraft industry, devising more efficient wings. Robert T. Jones arrived at Ames in August 1946 after distinguishing himself at Langley as the American inventor of the swept wing. Jones was a self-taught mathematician with a flair for aerodynamics. He became a protégé of theoretical aerodynamicist Max Munk and often claimed he was only extending Munk's ideas, though the clarity with which he expressed those ideas convinced everyone at Ames that Jones was his own genius. With his work on low aspect ratio wings, for example, Jones continued to show that the shapes of wings to come were far more than the assembly of airfoil sections—as NACA work at Langley had long ago proved. In jet aircraft, airfoil shapes blended into a new conception of the whole lifting surface—planform, sweep, aspect and aeroelasticity, all interacting in complex ways.

What Jones brought to the distinguished group of theorists at Ames—including Max Heaslet, Harvard Lomax, Milton Van Dyke and John Spreiter—was an intuitive feel for the importance of Mach cones (that is, the shock waves that spread like a cone back from the front of an aircraft). Ames had already begun studies on planforms that looked like arrowheads—long and slender with the leading edges swept back as much as 63 degrees.

A Culture of Research Excellence: 1939 – 1958

Atmosphere of Freedom

B-58 model, showing the fuselage pinched for the supersonic area rule, as well as conical camber in the wings.

Jones encouraged even more dramatic sweep, to 80 degrees, then devised theory supporting tests on triangular planforms. For example, Elliot Katzen led tests in the 1 by 3 foot supersonic tunnel, in 1955 and 1956, to determine which arrowhead shapes had the best possible lift-to-drag ratios at Mach 3, the cruise speed expected for a planned supersonic transport. Katzen had already designed five wings, using linear theory, with a similar arrowhead planform but differing twists and cambers. Jones consulted on the project, suggesting a planform swept back far enough behind the Mach cone so that the Mach number perpendicular to the leading edge was similar to that of the Boeing 707 in flight. He also suggested a Clark-Y airfoil with camber but no twist. When the thin metal model arrived from the model shop, Jones twisted the tips by hand until it looked right to him. This wing returned a lift-drag ratio of nine—the best efficiency ever measured for a wing travelling at Mach 3.

In 1952, Jones looked over the theory of the transonic area rule, which designers used to reduce the sharp drag rise at transonic speeds by controlling the simple cross section of the aircraft. Jones quickly devised the supersonic area rule, which led to designs that reduced drag at supersonic speeds by controlling the cross section of the aircraft cut by Mach cones. The big advantage of Jones' approach was that it was readily applicable to complete aircraft, including those carrying external weapon stores or fuel tanks.

Early in 1949, the research staff of Ames' 6 by 6 foot supersonic tunnel—Charles Hall, John Heitmeyer, Eugene Migotsky and John Boyd—concluded that dramatically new wing designs were needed to make jet aircraft operate at top efficiency. Theoretical analysis pointed them to a special form of camber—a slight convex curve—small at the root but increasing in depth and width toward the wing tip like the surface of a conical section. Experiments begun in 1950

John W. Boyd in 1955 explaining the efficiencies of conical camber.

confirmed their theoretical predictions of more uniform loading along the span. In 1953, at the early stages of its design, the Air Force asked Ames to study the disappointing efficiency of the Convair B-58 Hustler supersonic bomber. Hall's group designed a wing with conical camber that dramatically improved the range of the B-58, which in turn pioneered the design of all future supersonic transports. Likewise, the first Convair F-102 Delta Dagger was flown without conical camber in the wings. Ames showed that conical camber gave it an enormous improvement in range without diminishing its speed. Camber was built into all subsequent versions of the F-102. Overall, Ames tested 29 different aircraft to measure the improvements provided by conical camber.

The fastest jet wings were also the smallest, but small wings had trouble providing lift at slow speeds. Aerodynamicists at Ames refined the old ideas of boundary-layer control by applying suction or blowing to delay stall and give higher lift coefficients at low speeds during take-off and landing. Using the 7 by 10 foot tunnels, Ames researchers classified three types of stall encountered by airfoils, leading to better high-lift devices for aircraft. Using the 40 by 80 foot tunnel, Charles "Bill" Harper and John DeYoung tested the validity of the idea on entire aircraft. Using an old F-86 and mechanical techniques devised during Ames' earlier work on thermal deicing, Woody Cook, Seth Anderson and George Cooper collected data for a landmark study on flap suction.

The staff of Ames' 1 by 3 foot supersonic tunnel, who did more basic research as more development testing was moved to the 6 by 6 foot tunnel, led research into viscous flows. Dean Chapman, of the 1 by 3 foot supersonic tunnel branch, started work in 1947 on the effects of viscosity on drag at supersonic speeds, then returned to the California Institute of Technology to write up these data as his doctoral dissertation. His theoretical work led him to predict

A low aspect ratio wing model mounted on a sting in the 14 foot wind tunnel.

that blunt trailing edges worked better than sharp edges in minimizing viscous flows, which he verified experimentally. Those experiments generated data on base pressures that provided tools that designers could use to optimize the shapes of the back ends of aircraft—the aft part of the fuselage and trailing edges of the airfoil. In 1949, Chapman published a simplifying assumption, on the relationship between temperature and viscosity, that allowed better calculations of laminar mixing profiles and boundary layers. Chapman then continued his work on boundary layers, and supervised work that led to measurements of turbulent skin friction at Mach numbers up to 9.9 and at very high Reynolds numbers. He reached these high Mach and Reynolds numbers by constructing a boundary layer channel using high pressure helium as the test fluid. He then developed an equivalence relationship between helium measurements and air values. He matched those measurements with determinants of skin friction made at low Mach numbers by Donald Smith working in the 12 foot variable density tunnel.

Dynamic stability—that is, constantly changing relationships between the axes of motion—arose as another issue of high speed flight. At subsonic speeds, static stability of the aircraft could be easily checked during

Vortex generators mounted on the wing of a North American YF-86D test aircraft, to study how well they eliminated aileron buzz and buffeting in straight wing aircraft.

The NACA submerged inlet during comparison tests of a scoop inlet on the North American YF-93A. The YF-93A's were the first aircraft to use flush NACA engine inlets.

tunnel tests. Jet aircraft, however, had entirely new shapes, aerodynamic coefficients and mass distributions. Testing dynamic stability on jets was complicated work. Murray Tobak, working in the 6 by 6 foot wind tunnel, calculated the aerodynamic moments acting on models while they were rotating or oscillating about various axes. Benjamin Beam developed a technique for mounting models on springs, imparting an oscillation, then measuring these dynamic aerodynamic moments. He simplified the data processing by building an analog computer into the strain gauge circuitry. With this apparatus, Ames tested the dynamic stability of every new military aircraft in the 1950s.

Ames also addressed the complex airflow through jet air inlets. Jet turbine engines required much larger volumes of air than reciprocating engines, while also being more sensitive to the speed and turbulence of that air. The first jet aircraft inducted air through the nose, in part because the designers could rely on a wealth of NACA data on cowlings for reciprocating engines. When designers needed the nose for armament or radar, air intake scoops were moved back and submerged, following a design suggested by Charles Frick and Emmet Mossman working in the Ames 7 by 10 foot tunnels. Inlet design remained simple so long as jet aircraft remained subsonic.

For supersonic inlets, designers needed entirely new design principles and practices. Ames played a role in the design and testing of inlets for every early supersonic jet. Ames learned much from its work on the McDonnell F-101 Voodoo, which had been designed for subsonic flight until a better engine made supersonic flight possible. Ames quickly discovered what made inlets transition smoothly from the subsonic to the supersonic regime—attention to boundary layer removal, internal duct contours, and planned interaction between boundary layers and

Facing page: Model of the Unitary plan wind tunnel: (A) dry air storage spheres; (B) aftercooler; (C) 3 stage axial flow fan; (D) drive motors; (E) flow diversion valve; (F) 8 by 7 foot supersonic test section; (G) cooling tower; (H) flow diversion valve; (I) aftercooler; (J) 11 stage axial flow compressor; (K) 9 by 7 foot supersonic test section; (L) 11 by 11 foot transonic test section.

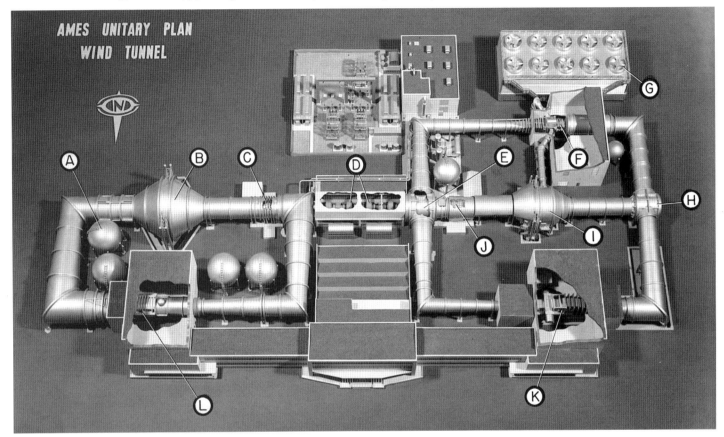

shock waves. Mossman devised a variable throat area that allowed for proper operations at any speed, and others at Ames continued their basic research into internal shock waves. As speeds approached Mach 2, jet designers started to use Ames data on supersonic compression within the duct.

UNITARY PLAN WIND TUNNEL

At the close of World War II, American aerodynamicists reflected on where they stood. They were surprised at how well British aerodynamicists had performed with limited resources. But they were amazed when they finally saw what German scientists had been working on—like jet propulsion and supersonic guided missiles—and concerned that a good many German scientists were now hard at work for the Soviet Union. American aerodynamicists felt that their pool of basic research had been exhausted while they solved urgent wartime problems. The NACA and the U.S. War Department independently decided that America needed to address the dawn of supersonic flight with more than fragments of theories and small scale tests.

NACA and military officials met in April 1946 and agreed on a "unitary plan"

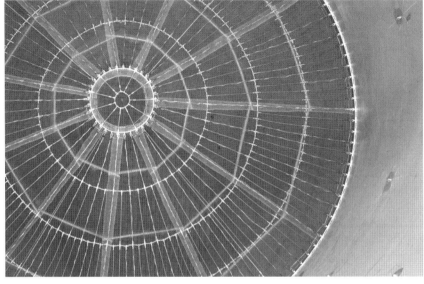

Safety screen in the diffuser of the Unitary plan wind tunnels.

for new facilities. They asked NACA member Arthur Raymond to head a Special Panel on Supersonic Laboratories, with members from the NACA, the Army, the Navy, airframe companies and engine companies. The Raymond panel report led to a new NACA special committee on supersonic facilities headed by Jerome Hunsaker. In January 1947, the Hunsaker Committee submitted its unitary plan, that was then scaled back by the U.S. joint research and development board to the most urgent facilities. The NACA got permission from the Bureau of the Budget to submit Unitary plan legislation to the House and Senate. It passed and was signed into law by President Truman on 27 October 1949. Under Title 1 of the Unitary Wind Tunnel Plan Act of 1949, NACA was to get $136 million for construction of facilities.

Construction of the 3 stage compressor for the Unitary plan tunnels.

Atmosphere of Freedom

The 6 by 6 foot wind tunnel, in August 1949, showing the rotor blades of the compressor. This was one of the earliest applications of multistage axial flow compressors in wind tunnels.

Yet, when they first reviewed this budget on 29 June 1950, Congress halved the authorization to $75 million. Now, rather than including facilities for the newly formed Air Force, the Unitary plan would only serve the combined interests of the three NACA laboratories. A year later, Congress again halved the appropriation and Ames prepared to lead the effort alone.

NACA Director Hugh Dryden established a NACA-wide project office for the Unitary wind tunnels, headquartered at Ames and led first by Jack Parsons and soon thereafter by Ralph Huntsberger. Huntsberger solicited suggestions from Langley and the Lewis laboratory, to make one complex of supersonic tunnels embody the ambitions of a nationwide plan. Construction began in 1951 at a cost of $32 million. After a six month shakedown, starting in June 1955, the tunnel opened with a test of the inlet design for the aircraft that became the McDonnell F-4 Phantom, the first designed for cruising at Mach 2 and the first to be procured by each military service.

"Unitary," in addition to describing the tunnel's political aspirations, also described the integration in its basic design. The Unitary facility covered 11 acres and consumed an enormous amount of electricity. It embodied three large test sections, powered by two large axial flow compressors that drove air over the Mach range of 0.3 to 3.5. A three-stage compressor drove air into a transonic section that was 11 by 11 feet, and an 11-stage compressor forced air through a rotating flow diversion valve into two supersonic sections that were 9 by 7 feet and 8 by 7 feet, respectively. Significantly, the speeds of the three test sections overlapped so that a single model could be tested over this entire range.

Each component of the Unitary pushed the state of the art in wind tunnel design. It embodied the largest diversion valves ever built, at 20 and 24 feet in diameter. Each weighed 250 tons, could be rotated in 25 minutes, and was airtight. The compressors were built by the Newport News Company. The rotor for the smaller compressor was, as of 1955, the largest cargo ever received at the Port of Oakland. The compressors were powered by

Northrop P-61 Black Widow with a recovery body model mounted below for a drop test to obtain transonic data.

four intercoupled motors built by General Electric, which were then the largest wound-rotor induction motors ever built. They were tandemly coupled between the two compressors, so that within thirty minutes, the motors could be disconnected from one compressor and connected to the other. Two of the motors had electrodynamic braking to slow the inertia of the compressors in case of emergency. The shafts carried the largest load of any tunnel shaft in the world. To reduce bending stress on the shafts, the entire drive train was supported by a single foundation.

The tunnel shell was a pressure vessel, constructed of steel up to 2.5 inches thick. Each test section had a separate nozzle configuration to match the Mach number required. The 11 foot tunnel had a simplified design, using a single jacking station to deflect a variable moment-of-inertia plate. A bypass valve equalized pressures between sections while a make-up air system controlled the temperature and humidity of the tunnel air by using intercoolers, dry air storage tanks, and evacuators. Dried air was pumped into storage tanks in a volume equal to that of the tunnel while humid air was evacuated. This controlled humidity to 100 parts per million of water, controlled stagnation pressure to 0.1 to 2.0 atmospheres, and greatly improved the attained Reynolds numbers.

The most important aircraft of the 1950s and 1960s were tested in the

Termination of a drop test in August 1950 to measure drag and pressure recovery on flush inlets.

A Culture of Research Excellence: 1939 – 1958

Unitary. In addition, Ames researchers explored the basic problems of the boundary layer, the mechanism of transition from laminar to turbulent flow, and the dynamic stability of various shapes used for warheads on ballistic missiles. Over the next four decades, the Unitary remained in almost constant use solving the evolving problems of supersonic flight. (In May 1996, it was dedicated as an International Historic Mechanical Engineering Landmark by the American Society of Mechanical Engineers.)

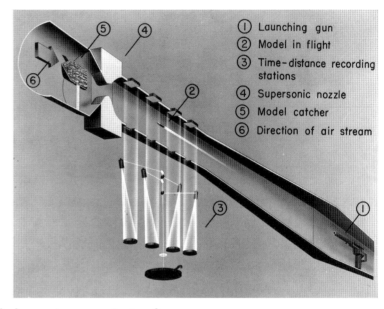

Schematic drawing of the supersonic free flight facility.

① Launching gun
② Model in flight
③ Time-distance recording stations
④ Supersonic nozzle
⑤ Model catcher
⑥ Direction of air stream

TRANSONICS

It is, perhaps, a testament to the experimental facilities at Ames that theory lagged far behind empirical advances in supersonic aircraft. In theory, it might seem to be no harder to theorize about the aerodynamic properties of bodies at transonic speeds—the speed range near Mach 1—than it is at subsonic or supersonic speeds. Yet prior to the late 1940s, nature revealed no solutions to either the theoretician or the experimenter. As a monument to nature's reluctance, there was a great store of experimental data that terminated at some Mach number close below 1, or started close above Mach 1. Furthermore, there were many theoretical predictions that simply did not agree with any experimental observations. Two developments in the late 1940s started to bring unity to the data above and below Mach 1. First was the development, by John Stack and his colleagues at Langley, of transonic tunnels with slightly open walls. Second was the small disturbance theory of transonic flow, advanced by work at Ames.

To move the calculations on small disturbance to the next level of approximation, transonic theory for two-dimensional flow required solution of a difficult nonlinear partial differential equation of a mixed elliptic-hyperbolic type. Walter Vincenti in the 1 by 3 foot tunnel attacked this problem using the hodograph method—a concept that had been explored by the Italian mathematician Tricomi in the 1920s—that transformed an intractable nonlinear equation into a more manageable linear equation. John Spreiter at Ames then summarized the basic equations needed for a useful approximation for Mach numbers nearly equal to unity. This allowed for fairly accurate prediction of transonic flows past very thin wings and slender bodies.

Max Heaslet, one of the few people at Ames to have a Ph.D., in mathematics, led the laboratory's theoretical aerodynamics section from 1945 to 1958, which did almost all the theoretical work that was not otherwise done separately by R.T. Jones. Heaslet's section undertook the systematic study of wing planforms for supersonic flight, and produced some exhaustive theoretical research on suggested wings in both steady and unsteady flows. This work on planforms was complemented by Spreiter's similarity laws and the forward and reverse flow theorems advanced by Heaslet and Spreiter and by Jones. Heaslet and Harvard Lomax, coupled with independent work by R.T. Jones, developed practical applications of theories of wing-body interference arising from the transonic area rule.

Milton Van Dyke developed a similar theoretical foundation for hypersonic flight. In 1954, he published the first-order small disturbance hypersonic equations useful as a guide in designing thin wings and bodies. Assisted by Helen Gordon, Van Dyke undertook the prediction of flow around the front of blunt-nosed missiles, an analysis so complex that he and Gordon relied upon electronic calculating machines. Alfred Eggers led another group that applied to hypersonic speeds the classic shock wave and expansion equations for supersonic flows. The criteria for applying these equations was the exact opposite of the small-perturbation methods, namely that the flow disturbance created by the body would be large. This generalized shock-expansion method was shown to allow rapid computation of a variety of hypersonic flows. Clarence Syvertson and David Dennis then improved the

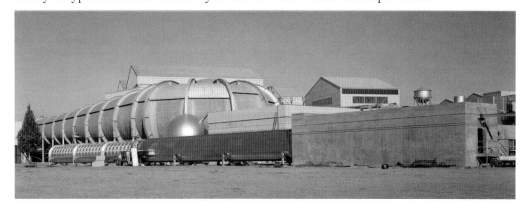

The pressurized ballistic range, in August 1957, housed in a long thin building near the 12 foot pressure tunnel.

Atmosphere of Freedom

Model, sabot, and cartridge—assembled and ready for firing in the supersonic free flight tunnel.

equations to develop a second-order shock expansion method for three-dimensional bodies. Flight in the hypersonic regime, because of work done at Ames, would have a firmer theoretical foundation as tunnel and flight tests began. Plus, Ames had shown how researchers with different skills and interests, concerned with separate but related issues, could calculate ever-better approximations of how real objects would move through real air.

HYPERSONICS: STEPPING UP TO THE SPACE AGE

Aerodynamicists still debate where to put the precise border between supersonic and hypersonic flight. Unlike the sharp jolt as a shock wave wraps around an aircraft near Mach 1, aircraft move gradually from the supersonic to hypersonic regime. Generally, hypersonic flight starts when the bow shock wave wraps closely around the vehicle and this shock wave generates heat high enough that air molecules vibrate, dissociate, and radiate heat and light, which heats up the aircraft structure. Chemical thermodynamics, thus, is as important in hypersonic design as aerodynamics. This heating generally starts at Mach 5 to Mach 10, or at speeds of one to two miles per second. In retrospect, these speeds had obvious importance for design of intercontinental missiles, satellites and reentry bodies. When Ames started its work, just after the war, chemical thermodynamics was an area of intense theoretical interest that Harvey Allen wanted to trailblaze.

Bringing hypersonic speeds to laboratory research required a stroke of ingenuity. In 1946, Allen suggested firing

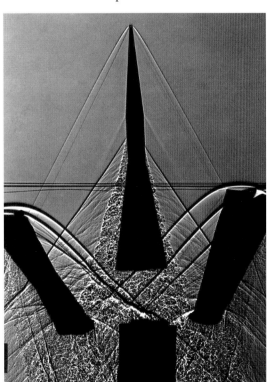

Shadowgraph of the model and sabot separating.

a model from a gun through the test section of a small supersonic tunnel. Thus, the speed of the model and the speed of the oncoming air combined to produce a hypersonic speed. Alvin Seiff took up the challenge of designing what came to be called, on its opening in 1948, the supersonic free flight tunnel (SSFFT). The engineering details to be worked out were immense, and challenged every branch of the Ames technical services division. Model shop craftsmen had to build tiny models, no bigger than a .22 caliber bullet, yet sturdy enough to be jolted into supersonic flight. The instrumentation branch had to obtain data from these models in free flight by rigging the tunnel with a series of very fast cameras and lights.

The full impact of this facility would be known a decade later during the human space missions, but its early use resulted in some important discoveries. The Ames high speed research division discovered an effect of skin-friction drag on turbulent boundary layers that had completely escaped notice in wind tunnel tests. In wind tunnel tests, the models were warmed to the temperature of the test air. In the free flight tests, the model skins were cold compared to the air—a condition comparable to actual flight—resulting in skin friction that was 40 percent greater than measured in tunnel tests. Simon Sommer and Barbara Short used these data to establish a formula to calculate the skin friction of turbulent boundary layers for a realistic range of Mach numbers and temperature conditions.

Another issue that was resolved was the theoretically and practically intriguing one of the transition of boundary layers from laminar to turbulent flow. Since laminar flows conduct less heat and cause less drag than the eddying flow of a turbulent layer, knowing where on a body the transition occurs is important in predicting heating and drag. The supersonic free flight facility was ideal for these studies. In addition to the comparable temperature conditions, turbulence in the free air stream was relatively low since much of the speed was contributed by

Transition from laminar to turbulent flow in the boundary layer of a missile at Mach 3.

LAMINAR   TRANSITION   TURBULENT

Ames from the sky in March 1958, shortly before it became part of NASA.

model motion. Plus, the shadowgraph cameras along the test section took excellent photographs of the state of the boundary layer. What the Ames group discovered was that the transition was unsteady, varying with time, and based on the model's angle of attack. From this, Allen and his group experimentally validated the importance of entry angle in designing missile warheads for laminar or turbulent flow.

Alex Charters took up the challenge of devising better guns to propel the free flight models ever faster. In 1952, he designed a gun using controlled explosions of light gas that could propel a test model faster than 14,000 feet per second—two times faster than standard powder guns. Once Charters constructed a prototype of his light-gas gun, DeFrance authorized construction of a hypervelocity ballistic range with a 600 foot long instrumented test range. Based on a challenge from Harvey Allen, in 1956 John Dimeff and William Kerwin of the Ames instrument development branch built a small model containing a calorimeter with a very simple telemetering circuit. Shakeout tests showed that this device could measure the heat transferred in free flight with great accuracy. Ames could now measure the temperature environment of the sensitive electronic components in the nose cones of guided missiles. When Ames opened its hypervelocity ballistic range in September 1957, it was used almost exclusively for development tests of guided missiles.

PREPARING FOR THE SPACE RACE

Ames' work in guided missiles and hypersonics put it in position to play a vital role in the missile race that dominated the aerospace industries around the world in the late 1950s. Ames' labor quota and budget got a short boost during the mini-mobilization surrounding the Korean War in the early 1950s. American military aircraft were then more consistently breaking the sound barrier, oftentimes in combat, which exposed new problems that Ames aerodynamicists were asked to solve. Once the Korean War ended, funding at Ames dropped. In 1953, its labor quota was 1,120, lower than in 1949. Furthermore, because of stagnation in civil service pay rates, DeFrance and Parsons were unable to fill many of the available positions. Soon, Ames would lose even more valuable employees to the higher wages in the aerospace industry.

In 1955, President Eisenhower declared that his top priorities would be two intercontinental ballistic missile projects—the Atlas and Titan—and three intermediate-range ballistic missiles—Thor, Jupiter and Polaris. Adjacent to Moffett Field, and far from its tradition-bound facilities in southern California, the new Lockheed Missiles and Space Company built its campus, including a great many clean rooms, in which they would construct Polaris missiles for the Navy. Lockheed also built a basic research laboratory that was one of the first tenants in the Stanford Research Park. While much of the work Lockheed did depended on access to the area's fast growing electronics industry, the company also hired many skilled workers away from Ames. Only civil service salary reforms, following the launch of the Soviet Sputnik in 1957, allowed DeFrance and Parsons to stem the flow.

Work poured into Ames as every branch of the military wanted help in designing and understanding its increasingly high-powered missiles. The NACA had embarked on a program to send experimental aircraft to ever-higher altitudes. The Bell X-2 experimental aircraft had already reached an altitude of 126,000 feet by 1956, and the hypersonic X-15 was expected to fly twice as high. Ames' budget soared too, from 1955 to 1958, as the NACA worked on making better missiles. Ames soon fell off this trajectory with the jolts, first, from the reconfiguration of the NACA into NASA and, second, the orientation of NASA around landing a human on the Moon.

The X-15 launches away from a B-52 with its rocket engine ignited.

Harvey Allen, chief of Ames' high speed research division explaining the blunt body concept.

1959   1968

TRANSITION INTO NASA

Chapter 2:
From a Laboratory to a Research Center

Ames contributed much of the technology that helped NASA succeed in the mission that most preoccupied it during the 1960s—sending an American to the Moon and returning him safely to Earth. Ames people defined the shape, aerodynamics, trajectory and ablative heat shield of the reentry capsule. They mapped out navigation systems, designed simulators for astronaut training, built magnetometers to explore the landing sites, and analyzed the lunar samples brought back. Still, compared with how it fueled growth at other NASA Centers, the rush to Apollo largely passed Ames by.

Ames' slow transition out of the NACA culture and into the NASA way of doing things, in retrospect, was a blessing. Under the continuing direction of Smith DeFrance, then Harvey Allen, Ames people quietly deepened their expertise in aerodynamics, thermodynamics, and simulation, then built new deep pockets of research expertise in the space and life sciences. They sat out the bureaucratic politics, feeding the frenzy toward ever more elaborate and expensive spacecraft. The gentle refocusing of Ames' NACA culture during the 1960s meant that Ames had nothing to unlearn when NASA faced its post-Apollo years—an era of austerity, collaboration, spin-offs, and broad efforts to justify NASA's utility to the American public.

RELATIONS WITH NASA HEADQUARTERS

President Dwight Eisenhower signed the National Aeronautics and Space Act into law on 29 July 1958, and its immediate impact was felt mostly in redefining Ames' relations with its headquarters. The NACA was disbanded, and all its facilities incorporated into the new National Aeronautics and Space Administration (NASA) which formally opened for business on 1 October 1958. Eisenhower wanted someone in charge of NASA who would take bold leaps into space and he appointed as administrator T. Keith Glennan, then president of the Case Institute of Technology. Hugh Dryden, who had been NACA chairman, was appointed Glennan's deputy. Glennan first renamed the three NACA "Laboratories" as "Centers," but kept Smith DeFrance firmly in charge of the NASA Ames Research Center.

HICONTA simulator (for height control test apparatus), in February 1969, mounted to the exterior framing of the 40 by 80 foot wind tunnel. It provided extraordinary vertical motion.

DeFrance went a year without making any organizational changes to reflect NASA's new space goals. At the end of 1959, he announced that Harvey Allen was promoted to assistant director, parallel to Russell Robinson. Robinson continued to manage most of Ames' wind tunnels, some of which were mothballed or consolidated into fewer branches to free up engineering talent to build newer tunnels. Allen's theoretical and applied research division was reconfigured so that he now managed an aerothermodynamics division and a newly established vehicle environment division. In addition, DeFrance formed an elite Ames manned satellite team, led first by Alfred Eggers and later by Alvin Seiff, that helped define the human lunar mission that would soon become NASA's organizational mission.

Perhaps the biggest cultural change at Ames came from personnel shifts. NASA

Ames Research Center, 14 December 1965.

also inherited the various space project offices managed by the Naval Research Laboratory—specifically Project Vanguard, upper atmosphere sounding rockets, and the scientific satellites for the International Geophysical Year. These offices had been scattered around the Washington, D.C. area, and Glennan decided to combine them at the newly built Goddard Space Flight Center in Beltsville, Maryland. Goddard would also be responsible for building spacecraft and payloads for scientific investigations, and for building a global tracking and data acquisition network. Glennan asked Harry Goett, chief of Ames' full scale and flight research division, to direct the new Goddard Center. Goett's departure, in August 1959, was a big loss for Ames. To replace Goett, DeFrance turned to Charles W. "Bill" Harper. Fortunately, Goett resisted the temptation to cannibalize colleagues from his former division, and instead built strong collaborative ties between Ames and Goddard, especially in the burgeoning field of space sciences.

The flood of money that started flowing through NASA only slowly reached Ames. The NACA budget was $340 million in fiscal 1959. As NASA, its budget rose to $500 million in fiscal 1960, to $965 million in fiscal 1961, and earmarked as $1,100 million for fiscal 1962. Staff had essentially doubled in this period, from the 8,000 inherited from the NACA to 16,000 at the end of 1960. However, most of this increase went to the new Centers—at Cape Canaveral, Houston, Goddard and Huntsville—and to the fabrication of launch vehicles and spacecraft. Ames people had little engineering experience in building or buying vehicles for space travel, even though they had devised much of the theory underlying them. Glennan, in addition, followed a practice from his days with the Atomic Energy Commission of expanding research and development

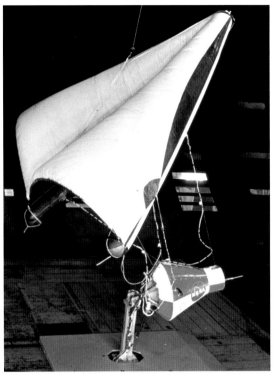

Model mounted in the 40 by 80 foot wind tunnel, for studies in 1962 on using paragliders to land space capsules.

through contracts with universities and industry rather than building expertise in-house. Thus, between 1958 and 1961, the Ames headcount dropped slightly to about 1,400, and its annual budget hovered around $20 million.

The disparity between what NASA got and what Ames received grew greater in early 1961 when President John Kennedy appointed James E. Webb to replace Glennan as administrator. Kennedy had campaigned on the issue of the missile gap and Eisenhower's willingness to let the Soviets win many "firsts" in space. So in Kennedy's second state of the union address, on 25 May 1961, he declared that by the end of the decade America would land an American on the Moon and return him safely to Earth. Ames people had already planned missions to the Moon and pioneered ways to return space travelers safely to Earth, but they had expected decades to pass before these plans were pursued. Kennedy's pronouncement dramatically accelerated their schedules. Kennedy immediately boosted NASA's fiscal 1962 budget by 60 percent to $1.8 billion and its fiscal 1963 budget to $3.5 billion. NASA's total headcount rose from 16,000 in 1960 to

Management process invaded Ames as the Center shifted from NACA to NASA oversight. Ames constructed a review room in its headquarters building where, in the graphical style that prevailed in the 1960s, Ames leadership could review progress against schedule, budget, and performance measures. Shown, in October 1965, is Merrill Mead, chief of Ames' program and resources office.

Steerable parachute for the Apollo capsule being tested in the 40 by 80 foot wind tunnel.

25,000 by 1963. More than half of this increase was spent on what Ames managers saw as the man-to-the-Moon space spectacular.

Again, Ames grew little relative to NASA, but it did grow. Ames' headcount less than doubled, from 1,400 in 1961 to 2,300 in 1965, while its budget quadrupled, from about $20 million to just over $80 million. Almost all of this budget increase, however, went to research and development contracts—thus marking the greatest change in the transition from NACA to NASA. Under the NACA, budgets grew slowly enough that research efforts could be planned in advance and personnel hired or trained in time to do the work. Under NASA, however, the only way to get skilled workers fast enough was to hire the firms that already employed them. Furthermore, under the NACA, Ames researchers collaborated with industrial engineers, university scientists, and military officers as peers who respected differences of opinions on technical matters. Under NASA, however, these same Ames researchers had enormous sums to give out, so their relations were influenced by money. Gradually,

Ames people found themselves spending more time managing their contractors and less time doing their own research.

Ames continued to report to what was essentially the old NACA headquarters group—guarded by Dryden, directed by Ira Abbott, and renamed the NASA Office of Advanced Research Programs. The four former NACA laboratories—Ames, Langley, Lewis, and the High Speed Flight Research Station—continued to coordinate their work through a series of technical committees. Even though the organizational commotion left in NASA's wake centered in the East, throughout the 1960s Ames found itself an increasingly smaller part of a much larger organization. Gradually the intimacy of the NACA organization faded as NASA's more impersonal style of management took over.

Four examples displayed the cultural chasm between Ames and the new NASA headquarters. First, in 1959 NASA headquarters told Ames to send all its aircraft south to Rogers Dry Lake—home of NASA's flight research station located at Edwards Air Force Base, California—except for those used in V/STOL research and one old F-86 used by Ames pilots to maintain their flight proficiency. Thus started decades of debate, and a series of subsequent

Schlieren photograph of a supersonic fighter aircraft model at Mach 1.4.

disagreements, over how aerodynamicists got access to aircraft for flight research. Second, NASA headquarters asserted its new right to claim for itself the 75.6 acres of Moffett Field on which Ames sat as well as 39.4 acres of adjacent privately held property. DeFrance argued that there was no need to change Ames' use permit agreement with the Navy, and he negotiated a support agreement that showed he was happy with Navy administration. Third, NASA renumbered the NACA report series but, more importantly, relaxed the restriction that research results by NASA employees first be published as NASA reports. New employees, especially in the space and life sciences, generally preferred to publish their work in disciplinary journals rather than through the peer networks so strong in the NACA days. Finally, NASA wanted Ames to leap into the limelight. DeFrance had encouraged Ames staff to shift any public attention to the sponsors of its research, and Ames' biggest outreach efforts had been the triennial inspections when industry leaders and local dignitaries—but no members of the public—could tour the laboratory. NASA headquarters encouraged DeFrance to hire a public information officer better able to engage general public audiences rather than technical or industry audiences. Bradford Evans arrived in August 1962 to lead those efforts, and soon Ames was hosting tours by local school groups.

General Dynamics F-111B aircraft, with its wings fully extended, undergoing tests in the 40 by 80 foot wind tunnel in 1969.

John Billingham, Melvin Sadoff, and Mark Patton of the Ames biotechnology division.

Ames moved more firmly into America's space program following three organizational changes. The first occurred in August 1962, when Harvey Allen formed a space sciences division and hired Charles P. Sonett to lead it. Sonett had worked for Space Technology Laboratories (later part of TRW, Inc.) building a variety of space probes for the Air Force, and he quickly established Ames as the leader in solar plasma studies.

The second organizational change was the start of life science research at Ames. Clark Randt had worked at NASA headquarters dreaming up biological experiments that could be carried into space. He decided that a laboratory was needed to do some ground experimentation prior to flight, and he thought Ames was a good place to start. So Randt sent Richard S. Young and Vance Oyama to work at Ames and build a small penthouse laboratory atop the instrument research building. Both reported back enthusiastically on how they were received. In the Bay Area, they had contact with some of the world's best biologists and physicians and, at Ames, they got help from a well-established human factors group in its flight simulation branch. With encouragement from headquarters, Ames established a life sciences directorate and, in November 1961, hired world-renowned neuropathologist Webb E. Haymaker to direct its many embryonic activities.

Lockheed JF-104A Starfighter piloted in 1959 by Fred Drinkwater to demonstrate very steep landing approaches of the type ultimately used with the space shuttle.

Shadowgraph of a flow field around a sharp nose cone at Mach 17.

Shadowgraph of a finned hemispherical body in free flight at Mach 2, during a 1958 test of the blunt body concept.

These life scientists, like the physical scientists that had long run Ames, were laboratory types who appreciated theory and its dependence upon experimentation. They, too, shunned operational ambitions. Yet these biologists still seemed grafted onto the Center. They used different disciplines, procedures and language. Many of the leading biologists were women, at a time when women were still sparse in the physical sciences. The biologists looked for success from different audiences, starting the fragmentation of the centerwide esprit de corps. Ames people had always been individualists, but all felt they were moving in the same general direction. Now, Ames served different intellectual communities and reorganized itself accordingly. Whereas Ames had always organized itself around research facilities, by 1963 it organized itself around disciplines throughout.

The third organizational change happened at headquarters. In November 1963, NASA headquarters reorganized itself so that Ames as a Center reported to the Office of Advanced Research and Technology (OART) while some major Ames programs reported to the other headquarters technical offices. DeFrance could no longer freely transfer money around the different programs at his Center. Headquarters staff had grown ten times since the NACA days, and from Ames perspective countless new people of uncertain position and vague authority were issuing orders. Some of these newcomers even bypassed the authority of the director and communicated directly with individual employees on budgetary and official matters. Virtually all of them wanted to know how Ames was going to help get a human on the Moon. Ames' NACA culture was under direct attack.

"…RETURNING HIM SAFELY TO EARTH"

By far the biggest contribution Ames made to NASA's human missions was solving the problem of getting astronauts safely back to Earth. Ames started working on safe reentry in 1951, when Harvey Allen had his eureka moment known as "the blunt body concept." In the early 1950s, while most attention focused on the rockets that would launch an object out of our atmosphere—an object like a nuclear-tipped ballistic missile—a few scientists started thinking about the far more difficult problem of getting it back into our atmosphere. Every known material would melt in the intense heat generated when the speeding warhead returned through ever-denser air. Most meteors burned up as they entered our atmosphere; how could humans design anything more sturdy than those? While many of the NACA's best aerodynamicists focused on aircraft to break the sound barrier, a few of its best

Model of the M-1 reentry body being mounted in the test throat of the 3.5 foot hypersonic tunnel.

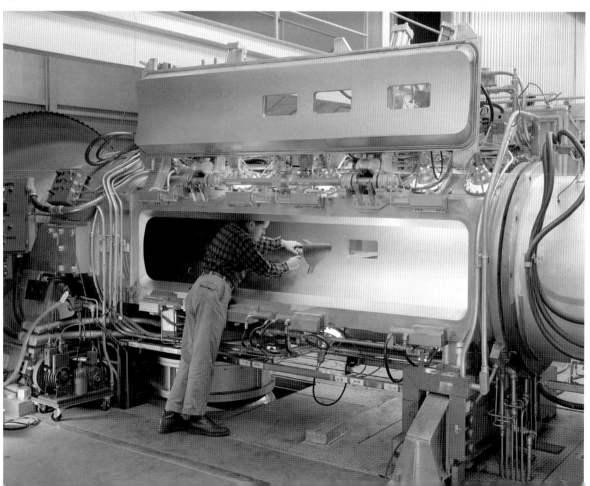

AMES 3.5-FOOT HYPERSONIC WIND TUNNEL

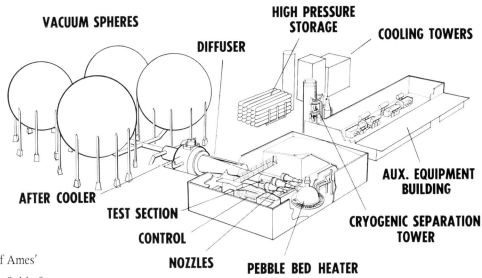

Schematic of the 3.5 foot hypersonic wind tunnel.

and brightest aerodynamicists focused instead on the thermal barrier.

Blunt Body Concept

H. Julian Allen and Alfred Eggers—working with Dean Chapman and the staff of Ames' fastest tunnels—pioneered the field of hypersonic aerodynamics. Though there is no clean dividing line between supersonics and hypersonics, most people put it between Mach 3 and 7 where heat issues (thermodynamics) become more important than airflow issues (aerodynamics). Allen and Eggers brought discipline to hypersonic reentry by simplifying the equations of motion to make possible parametric studies; by systematically varying vehicle mass, size, entry velocity and entry angle; and by coupling the motion equations to aerodynamic heating predictions. Allen soon came to realize that the key parameter was the shape of the reentry body.

A long, pointed cone made from heat-hardened metal was the shape most scientists thought would slip most easily back through the atmosphere. Less boundary layer friction meant less heat. But this shape also focused the heat on the tip of the cone. As the tip melted, the aerodynamics skewed and the cone tumbled. Allen looked at the boundary

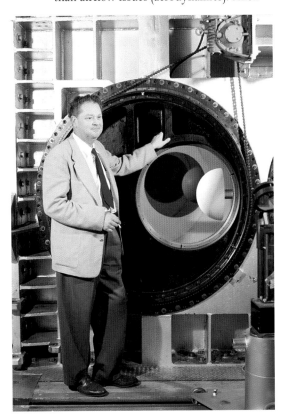

H. Julian Allen with a hemispheric model at the 8 by 7 foot test section of the Unitary plan tunnel.

Atmosphere entry simulator in 1958.

layer and shock wave in a completely different way. What if he devised a shape so that the bow shock wave passed heat into the atmospheric air at some distance from the reentry body? Could that same design also generate a boundary layer to carry friction heat around the body and leave it behind in a very hot wake? Allen first showed theoretically that, in almost all cases, the bow shock of a blunt body generated far less convective and friction heating than the pointy cone.

Allen had already designed a wind tunnel to prove his theory. In 1949, he had opened the first supersonic free flight facility—which fired a test model upstream into a rush of supersonic air—to test design concepts for guided missiles, intercontinental ballistic missiles and reentry vehicles. To provide ever better proof of his blunt body concept, Allen later presided over efforts by Ames researchers to develop light gas guns that would launch test models ever faster into atmospheres of different densities and chemical compositions.

Allen also showed that blunt reentry bodies—as they melted or sloughed off particles—had an important chemical interaction with their atmosphere. To explore the relation between the chemical structure and aerodynamic performance of blunt bodies, Ames hired and trained experts in material science. By the late 1950s, Ames researchers—led by Morris Rubesin, Constantine Pappas and John Howe—had pioneered theories on passive surface transpiration cooling (usually called ablation) that firmly moved blunt bodies from the theoretical to the practical. For example, Ames material scientists showed that by building blunt bodies from materials that gave off light gases under the intense heat of reentry, they could reduce both skin friction and aerodynamic heating.

Atmosphere of Freedom

Electric arc shock-tube facility, opened in 1966, was used to study the effects of radiation and ionization during outer planetary entries.

Meanwhile, Dean Chapman had developed a broad set of analytical tools to solve the problems of entry into planetary atmospheres, including calculations for the optimum trajectory to get a reentry body returning from the Moon back into Earth's atmosphere. Too steep an angle relative to the atmosphere, and the air about the body would get too dense too fast, causing the capsule to melt. Too shallow an angle, and the reentry capsule would skip off Earth's atmosphere like a flat rock on a smooth lake and continue off into space. First published in 1956, Chapman continued to refine his equations into the early 1960s. Hitting the precise trajectory angle that became known as the Chapman Corridor became the goal of navigation specialists elsewhere in NASA. At Ames, Chapman's methods were used to refine the aerodynamics of Allen's blunt body concept and define the thermodynamic envelope of the rarified atmosphere.

Ames applied its work on thermal structures, heating, and hypersonic aerodynamics to the X-15 experimental aircraft, which first flew faster than Mach 5 in June 1961 over Rogers Dry Lake. Data returned from the X-15 flight tests then supported modifications to theories about flight in near-space. But as America hurried its first plans to send humans into space and return them safely to Earth, NASA instructed Ames to make sure that every facet of this theory was right for the exact configuration of the space capsules. So in the early 1960s Ames opened several new facilities to test all facets—thermal and aerodynamic—of Allen's blunt body theory.

Hypervelocity Free Flight Facility

The hypervelocity research laboratory became the home of Ames' physics branch and carried out a significant body of research into ion beams and high temperature gases. Its 3.5 foot tunnel opened with interchangeable nozzles for operations at Mach 5, 7, 10 or 14. It included a pebble-bed heater which preheated the air to 3000 degrees Fahrenheit to prevent liquefaction in the test section at high Mach numbers. Ames added a 14 inch helium tunnel (at almost no cost) to the 3.5 foot tunnel building, which already had helium storage, and opened a separate 20 by 20 inch helium tunnel. These provided a very easy way of running preliminary hypervelocity tests from Mach 10 to Mach 25. Compared with air, helium allowed higher Mach numbers with the same linear velocities (feet per second). A one foot diameter hypervelocity shock tunnel, a remnant of the parabolic entry simulator, was built into an old Quonset hut. The shock tube could be filled with air of varying chemical composition, or any mixture of gases to simulate the atmosphere of Venus or Mars. It produced flows up to Mach 14, lasting as long as 100 milliseconds, with enthalpies up to 4000 Btu (British thermal units) per pound. Enthalpy indicated how much heat was transferred from the tunnel atmosphere to the tunnel model, and was thus a key measure in hypersonic research.

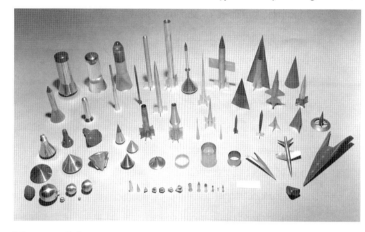

Models tested in the hypervelocity free flight tunnel.

The hypervelocity free flight facility (HFF), which grew out of this hypervelocity research laboratory, marked a major advance in Ames' ability to simulate the reentry of a body into an atmosphere. The idea of building a shock tunnel in counterflow with a light gas gun had been proven in 1958 with a small pilot HFF built by Thomas Canning and Alvin Seiff with spare parts. With a full-scale HFF budgeted at $5 million, Ames management wanted a bit more proof before investing so much in one facility. So in 1961, Canning and Seiff opened a 200 foot prototype HFF. Its two-stage shock compression gun hurled a projectile more than 20,000 feet per second into a shock tunnel that produced an air pulse travelling more than 15,000 feet per second. Ames had thus created a relative airspeed of 40,000 feet per second—the equivalent of reentry speed. Using this facility, Canning showed that

Hypersonic free flight gun, in June 1966, with Thomas Canning at the breech of the counterflow section.

Transition into NASA: 1959 – 1968

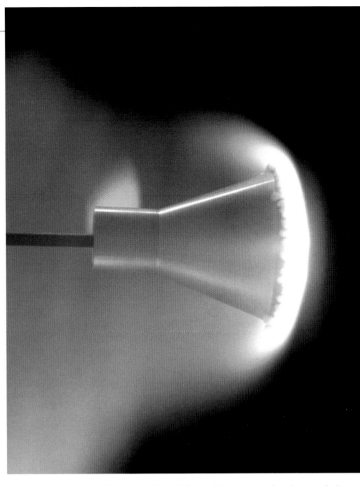

Ablation test of a Mercury capsule model.

the best shape for a space capsule—to retain a laminar boundary flow with low heat transfer— was a nearly flat face. Seiff also used it to test the flight stability of proposed capsule designs. Ames next increased the airspeed by rebuilding the piston driver with a deformable plastic that boosted the compression ratio. By July 1965, when the HFF officially opened, Ames could test models at relative velocities of 50,000 feet per second. To vary the Reynolds numbers of a test, Ames also built a pressurized ballistic range capable of pressures from 0.1 to 10 atmospheres. Every vehicle in America's human space program was tested there.

Arc Jets

While the HFF generated an enthalpy of 30,000 Btu per pound, the peak heating lasted mere milliseconds. These tunnels worked well for studying reentry aerodynamics, but the heating time was of little use for testing ablative materials. Ablative materials could be ceramics, quartz, teflon, or graphite composites that slowly melted and vaporized to move heat into the atmosphere rather than into the metal structure of the capsule. To test ablative materials—both how well they vaporized and how the melting affected their aerodynamics—Ames began developing the technology of arc jets. This work actually began in 1956, when Ames surveyed the state of commercial arc jets. Under pressure from NASA, in the early 1960s Ames designed its own. As the Apollo era dawned, Ames had a superb set of arc jets to complement its hypervelocity test facility.

These arc jets started with a supersonic blow-down tunnel, with air going from a pressurized vessel into a vacuum vessel. On its way through the supersonic throat the air was heated with a powerful electric arc—essentially, lightning controlled as it passed between two electrodes. The idea was simple but many problems had to be solved: air

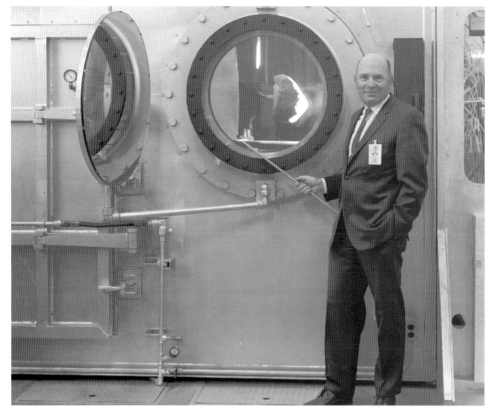

Glen Goodwin, chief of Ames' thermo and gas dynamics division, describing the workings of the broad plasma beam facility.

tends to avoid the electrical field of the arc so heating is not uniform; the intense heat melted nozzles and parts of the tunnel; and vaporized electrode materials contaminated the air.

So Ames devised electrodes of hollow, water-filled concentric rings, using a magnetic field to even out the arc. At low pressures, one of these concentric ring arc jets added to the airstream as much as 9000 Btu per pound of air. Though significant, this heating still did not represent spacecraft reentry conditions. Ames people looked for a better way of mixing the air with the arc. They devised a constricted arc that put one electrode upstream of the constricted tunnel and the other electrode downstream so that the arc passed through the narrow constriction along with the air. This produced enthalpies up to 12,000 Btu at seven atmospheres of pressure. By using the same constricted arc principle, but building a longer throat out of water-cooled washers of boron nitride, in late 1962 Ames achieved a supersonic arc plasma jet with enthalpies over 30,000 Btu per pound and heating that lasted several seconds. Expanding upon Ames' technical success in building arc jets, Glen Goodwin and Dean Chapman proposed a gas dynamics laboratory to explore how arc jets work in a comprehensive way. Opened in 1962, the $4 million facility accelerated the theoretical and empirical study of ablation.

By 1965, Ames had built a dozen arc jets to generate ever more sustained heat flows. An arc jet in the Mach 50 facility could operate with any mixture of gas, and achieved enthalpies up to 200,000 Btu per pound. As industrial firms began to design ablative materials for the Apollo heat shield, Ames researchers like John Lundell, Roy Wakefield and Nick Vojvodich could test them thoroughly and select the best.

Atmosphere of Freedom   Sixty Years at the NASA Ames Research Center

Apollo capsule free flight ablation test.

Impact Physics and Tektites

For clues on reentry aerodynamics, Allen also suggested that Ames study meteorites, nature's entry bodies. Using their high-speed guns, Ames first explored the theory of meteor impacts by hurling spheres of various densities at flat targets. At the highest impact speeds, both the sphere and target would melt and splash, forming a crater coated with the sphere material—very much like lunar craters. Ames then turned its attention to lunar craters—specifically the radial rays of ejected materials—by shooting meteor-like stones at sand targets like those on the Moon. By concluding that an enormous volume of material was ejected from the Moon with every meteor impact, they paved the way for lunar landings by suggesting that the surface of the Moon was most likely all settled dust.

One stunning example of what results when Ames' raw scientific genius is unleashed was the work of Dean Chapman on tektites. In early 1959, Chapman used the 1 by 3 foot blowdown tunnel (as it was about to be dismantled) to melt frozen glycerin in a Mach 3 airstream. In the frozen glycerin he first photographed the flattening of a sphere into a shape similar to Allen's blunt body. The ball quickly softened, its surface melted into a viscous fluid, and a system of surface waves appeared that were concentric around the aerodynamic stagnation point. On his

Gas dynamics facility, in 1964, and the 20 inch helium tunnel.

Impact test, simulating space debris hitting an orbiting capsule. The spark came from a blunt-nose, 20 millimeter polyethylene model hitting an aluminum target at 19,500 feet per second in a pressure simulated as 100,000 foot altitude.

Dean Chapman showing a tektite to Vice President Lyndon Johnson in October 1961.

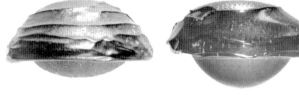

A natural tektite, at left, compared with an artifical tektite.

way to England for a year of research, Chapman visited a geologist at the American Museum of Natural History, who saw some similarity in the wave patterns on the glycerin balls and the wave patterns on glassy pellets of black glass called tektites. Tektites had been unearthed for centuries, mostly around Australia, though geologists still vigorously debated their origin. When geologists asked the Australian aborigines where the tektites came from, they pointed vaguely up to the sky.

Chapman applied the skills he had—in aerodynamics and ablation—and learned what chemistry he needed. He cut open some tektites and found flow lines that suggested they had been melted into button shapes, after having been previously melted into spheres. From the flow lines he also calculated the speed and angle at which they entered Earth's atmosphere. He then melted tektite-type material under those reentry conditions in Ames' arc jet tunnels. By making artificial tektites, he established that they got their shape from entering Earth's atmosphere just as a space capsule would.

Chapman next offered a theory of where the tektites came from. By eliminating every other possibility, he suggested that they came from the Moon. Ejected fast enough following a meteor impact, these molten spheres escaped the Moon's gravitational field, hardened in space, then were sucked in by Earth's gravitation. Harvey Allen walked into Chapman's office one day and egged him on: "If you're any good as a scientist you could tell me exactly which crater they came from." So Chapman accepted the challenge, calculated the relative positions of Earth and Moon, and postulated that they most likely came from the Rosse Ray of the crater Tycho.

Transition into NASA: 1959 – 1968

Atmosphere of Freedom   Sixty Years at the NASA Ames Research Center

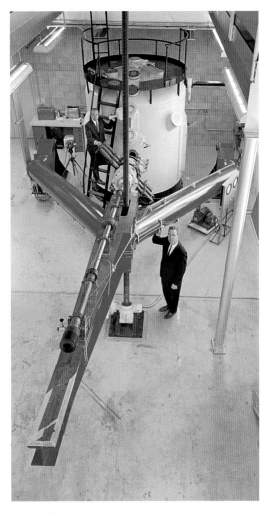

Thirty caliber vertical impact range, in 1964, with the gun in the horizontal loading position. William Quaide and Donald Gault of the Ames planetology branch used the gun range to study the formation of impact craters on the Moon.

In October 1963, Chapman won NASA's Medal for Exceptional Scientific Achievement. His bit of scientific sleuthing had accelerated curiosity about the composition of the Moon and the forces that shaped it, in the process validating some theories about ablation and aerodynamic stability of entry shapes. But the community of terrestrial geologists kept open the debate. While most geologists now accepted that tektites had entered Earth's atmosphere at melting speeds, most maintained that they were terrestrial in origin—ejected by volcanoes or a meteor crash near Antarctica. Only a single sample returned from the Moon, during Apollo 12, bears any chemical resemblance to the tektites. Thus, only the return of samples from the Rosse Ray would ultimately prove Chapman's theory of lunar origin.

FLIGHT STUDIES

Of course, not every aerodynamicist at Ames was working on the Apollo project. Ames continued working on high-speed aerodynamics, such as boundary layer transition, efficient supersonic inlets, dynamic loads on aircraft structures, and wing-tip vortices. Ames focused its work on high-lift devices to test new approaches to vertical and short take-off and landing aircraft. Ames continued to use its wind tunnels to clean up the designs of modern commercial aircraft as air passengers took to the skies in the new jumbo jets. And Ames solved many of the seemingly intractable flight problems of military aircraft—problems often uncovered during action in Vietnam.

Ames also continued to do airplane configuration studies, most notably for

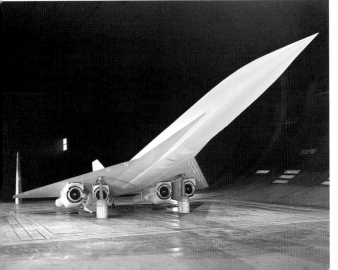

Double-delta planform on a supersonic transport model, mounted in the 40 by 80 foot wind tunnel.

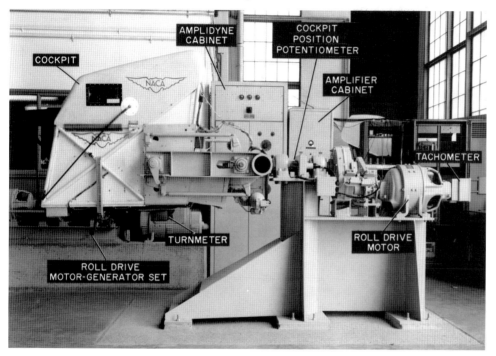

A simple pitch-roll chair, a 2-degree-of-freedom simulator built in 1958.

the supersonic transport. NASA decided it would outline the general configuration from which an aircraft firm would build a commercial supersonic transport (SST). Because of Ames' long interest in delta wings and canards—going back to tests of the North American B-70 supersonic bomber—Victor Peterson and Loren Bright of Ames helped develop a delta-canard configuration. The Ames vehicle aerodynamics branch also suggested a double-delta configuration that Lockheed used for its SST proposal. Then Ames used its wind tunnels to help the Federal Aviation Administration (FAA) to evaluate the efficiency and environmental impact of the designs. And Ames used its flight simulators to coordinate handling qualities research by NASA, pilot groups, industrial engineers, and airworthiness authorities from the United States, the United Kingdom, and France. Ames thus led development of criteria used to certify civil supersonic transports; the European-built Concorde was certified to these criteria in both Europe and the United States.

Ames people are famous for reinventing themselves to apply the skills they have to problems that are just being defined. One example of personal reinvention, in the

The Ames 5-degree-of-freedom simulator, 1962.

Transition into NASA: 1959 – 1968

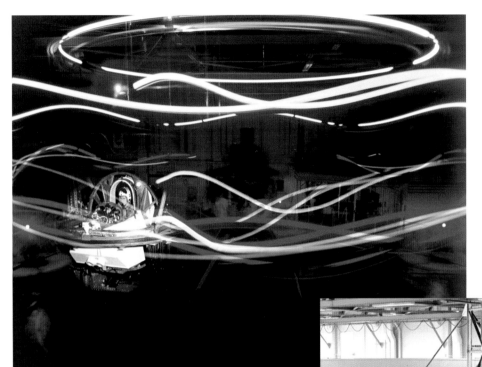

The 5-degree-of-freedom flight simulator, in 1962, with time-lapsed exposure to show its wide range of motion.

The 5-degree-of-freedom piloted flight simulator.

1960s, is Ames' emergence as a leader in flight simulators. Ames had begun building simulators in the early 1950s, when the Center acquired its first analog computers to solve dynamics, and as part of Ames' work in aircraft handling qualities. Harry Goett had pushed Ames to get further into simulator design, and George Rathert had led the effort. Ames' analog computing staff recognized that they could program the computer with an aircraft's aerodynamics and equations of motion, that a mockup of the pilot stick and pedals could provide computer inputs, and that computer output could drive mockups of aircraft instrumentation. Thus, the entire loop of flight control could be tested safely on the ground. Simulators for entry-level training were already widely used, but by building their system around a general, reprogrammable computer, Ames pioneered development of the flight research simulator.

By the late 1950s, using parts scrounged from other efforts, Ames had constructed a crude roll-pitch chair. Goett championed construction of another simulator, proudly displayed at the Ames 1958 inspection, to test design concepts for the X-15 hypersonic

The 6-degree-of-freedom motion simulator, opened in 1964, was used to investigate aircraft handling qualities, especially for takeoff and landing studies. The cab is normally covered, with visuals provided by a TV monitor.

experimental aircraft. Ames was ready to move when NASA asked for simulators to help plan for spacecraft to be piloted in the unfamiliar territory of microgravity. Fortunately, Ames had on staff a superb group of test pilots and mechanics who wanted to stay at Ames even after NASA headquarters sent away most of its aircraft. Led by John Dusterberry, this analog and flight simulator branch pioneered construction of sophisticated simulators to suit the research needs of other groups at Ames and around the world.

In 1959, Ames embarked on an ambitious effort to build a five-degree-of-freedom motion simulator. This was a simulated cockpit built on the end of a 30 foot long centrifuge arm, which provided curvilinear and vertical motion. The cockpit had electrical motors to move it about pitch, roll and yaw. It was a crude effort, built of borrowed parts by Ames' engineering services division. But the simulator proved the design principle, pilots thought it did a great job representing airplane flight, and it was put to immediate use on stability augmentors for supersonic transports.

In 1963, Ames opened a six-degree-of-freedom simulator for rotorcraft research, a moving cab simulator for transport aircraft, and a midcourse navigation simulator for use in training Apollo astronauts. Ames combined its various simulators into a spaceflight guidance research laboratory, opened in 1966 at a cost of $13 million. One of the most important additions was a centrifuge spaceflight

Transition into NASA: 1959 – 1968

Atmosphere of Freedom Sixty Years at the NASA Ames Research Center

Apollo navigation simulator, used to test concepts for midcourse correction on the voyage to and from the Moon.

simulator at the end of a centrifuge arm, capable of accelerating at a rate of 7.5 g forces per second. Another was a satellite attitude control facility, built inside a 22 foot diameter sphere to teach ground controllers how to stabilize robotic spacecraft.

Ames had become the best in the world at adding motion generators to flight simulators, and soon pioneered out-the-window scenes to make the simulation seem even more realistic for the pilot. Ames also emphasized the modular design of components, so that various computers, visual projectors, and motion systems could be easily interconnected to simulate some proposed aircraft design.

Ames also made key contributions to flight navigation. Stanley Schmidt had joined Ames in 1946, working in instrumentation, analog computing and linear perturbation theory. In 1959, when NASA first tasked its Centers to explore the problems of navigating to the Moon, Schmidt saw the potential for making major theoretical extensions to the Kalman linear

Brent Creer, chief of the Ames manned spacecraft simulation branch, developed the Apollo midcourse navigation and guidance simulator. Here he is shown with sextants designed to be carried aboard the capsule.

This human-carrying rotation device opened in 1966. It was used in studies of motion sickness, pilot response to microgravity, and in studies of pilot sensing of rotation.

filter. The result was a state-estimation algorithm called the Kalman-Schmidt filter. By early 1961, Schmidt and John White had demonstrated that a computer built with this filter, combined with optical measurements of the stars and data about the motion of the spacecraft, could provide the accuracy needed for a successful insertion into orbit around the Moon. Meanwhile Gerald Smith, also of the Ames theoretical guidance and control branch, demonstrated the value of ground-based guidance as a backup to guidance on board the Apollo capsules. The Kalman-Schmidt filter was embedded in the Apollo navigation computer and ultimately into all air navigation systems, and laid the foundation for Ames' future leadership in flight and air traffic research.

In the mid-1960s, Ames also participated in the design of suits for astronauts to wear for extravehicular activity. Though none of the concepts demonstrated by Ames were included in the Apollo spacesuits, many were incorporated in the next-generation of suits designed for Space Shuttle astronauts. Hubert "Vic" Vykukal led Ames' space human factors staff in designing the AX-1 and AX-2 suits for extended lunar operations, and in validating the concepts of the single-axis waist and rotary bearing joints. The AX-3 spacesuit was the first high pressure suit—able to operate at normal Earth atmospheric pressures—and demonstrated a low-leakage, low-torque bearing. Ames continued to advance spacesuit concepts well beyond the Apollo years, and some concepts were applied only two decades later. The AX-5 suit, designed for the International Space Station, was built entirely of aluminum with only fifteen major parts. It has stainless steel rotary

Vic Vykukal modeling the AX-1 spacesuit in 1966.

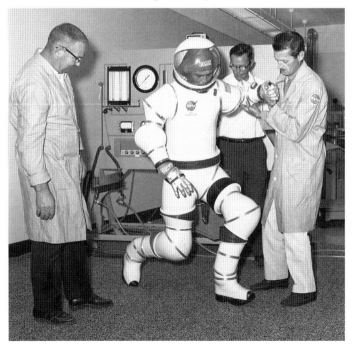

Transition into NASA: 1959 – 1968

bearings and no fabric or soft parts. The AX-5 size can be quickly changed, it is easy to maintain, and it has excellent protection against meteorites and other hazards. Because it has a constant volume, it operates at a constant internal pressure, so it is easy for the astronaut to move. Ames also developed a liquid cooled garment, a network of fine tubes worn against the skin to maintain the astronaut's temperature. To expedite Ames' efforts in spacesuit design, in September 1987 Ames would open a neutral buoyancy test facility, only the third human-rated underwater test facility in the country. In building these suits, as in building the simulators for aircraft and spaceflight, Ames came to rely upon experts in human physiology joining the Center's burgeoning work in the life sciences.

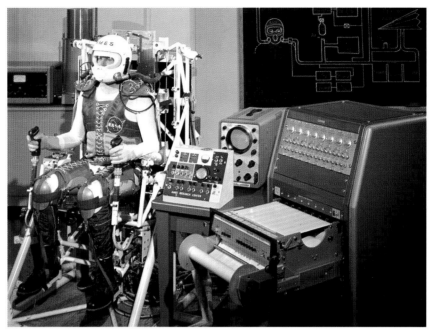

A 1962 study of breathing problems encountered during reentry, with pilot Robert St. John strapped into a respiratory restraint suit and a closed-loop breathing system.

START OF LIFE SCIENCES RESEARCH

In the early 1960s, as in the early 1940s, Ames looked like a construction zone. Not only were new arc jet and hypervelocity tunnels being built at top speed, but the life sciences division had to

Flight and guidance centrifuge in 1971 was used for spacecraft mission simulations and research on human response to motion stress.

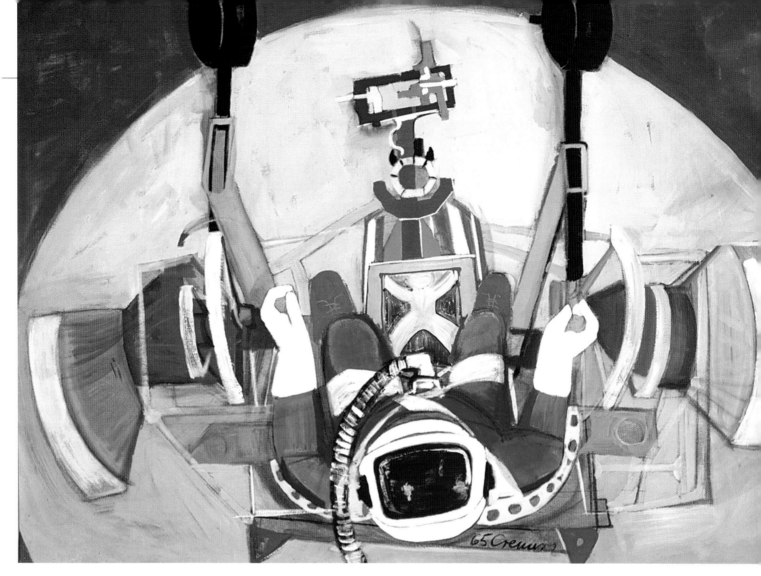

Artwork of an astronaut training for the Gemini missions using a simulator chair based on an Ames design.

build numerous facilities from scratch. The first biologists to move out of their temporary trailers, in 1964, moved into the biosciences laboratory. Much of this laboratory was an animal shelter, where Ames housed a well-constructed colony of several hundred pig-tail macaques from southeastern Asia for use in ground-based control experiments prior to the Biosatellite missions. In December 1965, Ames dedicated its life sciences research laboratory. It was architecturally significant within the Ames compound of square, two story, concrete-faced buildings, because it stood three stories tall and had a concrete surfacing dimple like the Moon. It cost more than $4 million to build and equip its state-of-the-art exobiology and enzyme laboratories.

These new facilities were designed to help Ames biologists understand the physiological stress that spaceflight and microgravity imposed on humans. While the Manned Spacecraft Center near Houston screened individual astronauts for adaptability and led their training, Ames developed the fundamental science underlying this tactical work. Mark Patton in the Ames biotechnology division studied the performance of humans under physiological and psychological stress to measure, for example, their ability to see and process visual signals. Other studies focused on how well humans adapted to

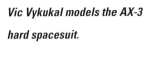

Vic Vykukal models the AX-3 hard spacesuit.

long-term confinement, what bed rest studies showed about muscle atrophy, and what sort of atmosphere was best for astronauts to breathe. Ames' growing collection of flight simulators also was used for fundamental studies of human adaptability to the gravitational stress of lift-off, microgravity in spaceflight, and the vibration and noise of reentry. All these data helped define the shape and function of the Gemini and Apollo capsules.

Ames' environmental biology division studied the effect of spaceflight on specific organs, mostly through animal models. Jiro Oyama pioneered the use of centrifuges to alter the gravitational environment of rats, plants, bacteria and other living organisms, and thus pioneered the field of gravitational biology. In conjunction with the University of California Radiation Laboratory, Ames used animal models to determine if the brain would be damaged by exposure to high-energy solar rays that are usually filtered out by Earth's atmosphere. To support all this life sciences research, Ames asked its instrumentation group to use the expertise it had earned in building sensors for aircraft to build bio-instrumentation. Under the guidance of John Dimeff, the Ames instrumentation branch built sophisticated sensors and clever telemetry devices to measure and record all sorts of physiological data.

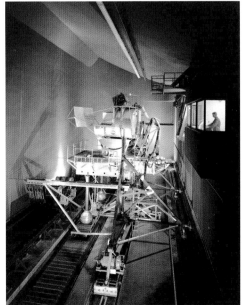

Flight Simulator for Advanced Aircraft (FSAA), opened in 1969, was used to investigate the landing, takeoff and handling qualities of large aircraft. The control room is on the right.

Building Blocks of Life

Exobiology, however, generated the most headlines during Ames' early work in the life sciences. As the task was first given to Ames, exobiology focused on how to identify any life encountered in outer space. Harold P. "Chuck" Klein had worked for eight years at Brandeis University defining what nonterrestrial life might look like in its chemical traces. He arrived at Ames in 1963 to head the exobiology branch and guided construction of Ames' superb collection of gas chromatographs, mass spectrometers, and quarantine facilities. A year later DeFrance asked Klein, who had served as chairman of Brandeis' biology department, to become director of Ames' life sciences directorate. Klein brought intellectual coherence to Ames' efforts, fought for both support and distance from Washington, and did a superb job recruiting scientists from academia.

Cyril Ponnamperuma of the Ames chemical evolution branch with the electrical-discharge apparatus used in his experiments on the chemical orgins of life.

Cyril Ponnamperuma arrived at Ames in the summer of 1961 in the first class of postdoctoral fellows under a joint program between NASA and the National Research Council. What he saw at Ames led him to join the permanent staff, and for the next decade he infused Ames' exobiology efforts with a flourish of intellectual energy. Using all that NASA scientists were learning about the chemical composition of the universe, Ponnamperuma brought a fresh outlook to the question of how life began at all.

Geologists had already discovered much about the chemical composition of primordial Earth. Scientists at Ames used their chromatographs and spectroscopes to detect the minute amounts of organic compounds in extraterrestrial bodies, like meteorites. From this, Ponnamperuma's colleagues in Ames' chemical evolution branch elucidated the inanimate building blocks and natural origins of life. Like many biochemists, they

The evolution of life on Earth, depicted from its chemical origins on the left to mammalian life on the right.

suspected that life was simply a property of matter in a certain state of organization, and if they could duplicate that organization in a test tube then they could make life appear. If they did, they would learn more about how to look for life elsewhere in the universe.

By the end of 1965, in apparatus designed to simulate primitive Earth conditions, Ponnamperuma and his group succeeded in synthesizing some of the components of the genetic chain—bases (adenine and guanine), sugars (ribose and deoxyribose), sugar-based combinations (adenosine and deoxyadenosine), nucleotides (like adenosine triphosphate), and some of the amino acids. A breakthrough came when the Murchison carbonaceous meteorite fell on Australia in September 1969. In the Murchison meteorite, Ames exobiologists unambiguously detected complex organic molecules—amino acids—which proved prebiotic chemical evolution. These amino acids were achiral (lacking handedness), thus unlike the chiral amino acids (with left handedness) produced by any living system. The carbon in these organic compounds had an isotope ratio that fell far outside the range of organic matter on Earth. The organic compounds in the Murchison meteorite arose in the parent body of the meteorite, which was subject to volcanic outgassing, weathering and clay production as occurred on prebiotic Earth.

Lunar Sample Analysis

Because of the expertise Ames people had developed in the chemical composition of nonterrestrial environments and in the

Apollo 12 lunar module over the lunar surface. Apollo 12 left an Ames magnetometer on the Moon as part of a package of scientific instruments.

life sciences, headquarters asked Ames to build one of two lunar sample receiving facilities. To prevent any contamination of the samples, this facility had to be very clean, even beyond the best of the Silicon Valley clean rooms. Whereas the facility at the Manned Spacecraft Center in Houston focused on identifying any harmful elements in the lunar samples, Ames scientists looked at the overall composition of the lunar regolith (the term for its rocky soil).

Ames researchers—led by Cyril Ponnamperuma, Vance Oyama and William Quaide—examined the carbon chemistry of the lunar soils, and concluded that it contained no life. But this conclusion opened new questions. Why was there no life? What kind of carbon chemistry occurs in the absence of life? Continuing their efforts, Ames researchers discovered that the lunar regolith was constantly bombarded by micrometeorites and the solar wind, and that interaction with the cosmic debris and solar atomic particles defined the chemical evolution of the surface of the Moon.

Ames also provided tools for investigating the chemistry of the Moon beneath its surface. Apollos 12, 14, 15, and 16 each carried a magnetometer—designed by Charles Sonnet, refined by Palmer Dyal, and built at Ames around an advanced ring core fluxgate sensor. These were left at the Apollo lunar landing sites to radio back data on the magnetic shape of the Moon. Paced by a stored program, these magnetometers first measured the permanent magnetic field generated by fossil magnetic materials. They then measured

Thr tri-axis magnetometer, developed at Ames, and used to measure magnetic fields on the Moon.

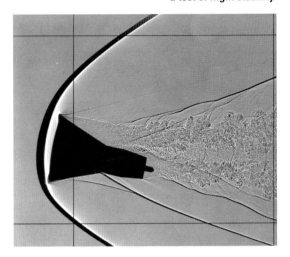

Shadowgraph of the Gemini capsule model in a test of flight stability.

the electrical conductivity and temperature profile of the lunar interior, from which scientists deduced the Moon's magnetic permeability and its iron content. And they measured the interactions of the lunar fields with the solar wind. For Apollos 15 and 16, Ames also developed handheld magnetometers to be carried aboard the lunar rover.

The magnetometer left on the Moon by Apollo 12 showed that the Moon does not have a two-pole magnetism as does Earth. It also suggested that the Moon is a solid, cold mass, without a hot core like that of Earth. But it also unveiled a magnetic anomaly 100 times stronger than the average magnetic field on the Moon. The series of magnetometers showed that the Moon's transient magnetic fields were induced by the solar wind and that they varied from place to place on the surface. Most important, these data allowed NASA to develop plans for a satellite to map in detail the permanent lunar magnetic fields in support of future missions to the Moon. These efforts in the space and life sciences displayed Ames' strengths in basic research and experimentation, but they were not at the heart of NASA's early missions.

SPACE PROGRAM MANAGEMENT

Smith DeFrance and Harvey Allen both insisted that Ames stick to research—either basic or applied—and stay out of what NASA called project management. Russ Robinson agreed, and so did Ira Abbott at NASA headquarters. Jack Parsons, though, encouraged the many young Ames researchers who wanted to try their hand at project management, and so did Harry Goett. Early in 1958, Goett and Robert Crane prepared specifications for a precise attitude stabilization system needed for the orbiting astronomical observatory (OAO), as well as the Nimbus meteorological satellite. Encouraged by how well NASA headquarters received their ideas, Goett convinced DeFrance to submit a proposal for Ames to assume total technical responsibility for the OAO project. Abbott, with Dryden's concurrence, told Ames to stick to its research.

Al Eggers, backed by the expertise pulled together in his new vehicle environment division, was the next to try to get Ames involved in project management. Eggers' assistant division chief, Charles Hall, wanted to build a solar probe. By late 1961, Hall had succeeded in getting two audiences with headquarters staff, who discouraged him by suggesting he redesign it as an interplanetary probe. Space Technology Laboratories (STL) heard of Ames' interest, and Hall was able to raise enough money to hire STL for a feasibility study of an interplanetary probe. Armed with the study, DeFrance and Parsons both went to headquarters and, in November 1963, won the right for Ames to manage the PIQSY probe (for Pioneer international quiet sun year), a name soon shortened to Pioneer.

DeFrance also reluctantly supported the Biosatellite program. Biosatellite started when headquarters asked Ames what science might come from sending monkeys into space in leftover Mercury capsules. When Carlton Bioletti submitted Ames' report to headquarters early in 1962, an intense jurisdictional dispute erupted with the Air Force over which agency should control aerospace human factors research. Because the United States was already well behind the Soviet Union in space life sciences, NASA won this battle and immediately established the life sciences directorate at Ames. In the meantime, biologists had started submitting unsolicited proposals to Ames. Bioletti and his small group of ten visited each of these biologists to sketch out the specifications for a series of biological satellites. Impressed with these efforts, in October 1962 Ames was tasked to manage Project Biosatellite.

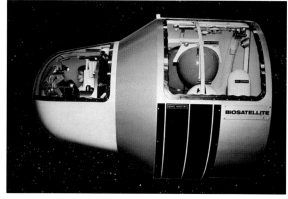

Biosatellite model with monkey shown in the front of the capsule and the life-support package in the rear.

Ames' work in lifting bodies also took it, slowly, into project management. Eggers and his group in the 10 by 14 inch tunnel in 1957 had conceived of a spacecraft that could safely reenter Earth's atmosphere, gain aerodynamic control and land like an airplane. They called these "lifting bodies" because the lift came from the fuselage rather than from wings, which were too vulnerable to melting during reentry. Using every tunnel available to them, Ames aerodynamicists formalized the design, tunnel tested it, and procured a

M2-F2 lifting body mounted in the 40 by 80 foot wind tunnel in July 1965 prior to flight tests.

flying prototype called the M2-F2 from Northrop for flight tests at NASA's High Speed Flight Station beginning in 1965. These tests, in conjunction with flight tests of the SV-5D and HL-10 lifting bodies, gave NASA the confidence it needed to choose a lifting body design for the Space Shuttle.

By 1963, even DeFrance had to recognize that without some experience in how projects were managed, Ames would be left behind NASA's growth curve. The NACA culture indicated that any scientist interested in a project should execute it. That had been possible even on the larger wind tunnels because a scientist only needed the help of Jack Parsons to marshal resources within the laboratory. When projects were launched into space, however, executing projects got substantially more complex. First, most of the support came from outside the Center—from aerospace contractors or from the NASA Centers that built launch vehicles, spacecraft, or data acquisition networks. Second, nothing could be allowed to go wrong when the spacecraft or experimental payload was so distant in space, so technical integration and reliability had to be very well-conceived and executed. Finally, the larger costs evoked greater suspicion from headquarters, and thus warranted more preliminary reporting on how things would go right. Scientists were increasingly willing to have a project

The M2-F2 lifting body returns from a test flight at the Dryden Flight Research Center with an F-104 flying chase. On its first flight on 12 July 1966 the M2-F2 was piloted by Milt Thompson. The M2-F2 was dropped from a wing mount on NASA's B-52 at an altitude of 45,000 feet. The M2-F2 weighed 4,620 pounds, was 22 feet long, and was 10 feet wide.

management specialist handle these more burdensome support arrangements.

Project management was the sort of integrative, multidisciplinary work that engineers excelled in, but spare engineers were hard to find at Ames. So Ames management began to cultivate some project managers attuned to the scientists they would serve. Bob Crane was named to the new position of assistant director for development and he, in turn, named John V. Foster to head his systems engineering division. Charlie Hall then managed the Pioneer project, and Charlie Wilson managed the Biosatellite. Both Hall and Wilson worked with lean staffs, who oversaw more extensive contracting than was usual at Ames. They studied NASA protocols for network scheduling and systems engineering. Significantly, both reported to headquarters through the Office of Space Science and Applications (OSSA), whereas the Center as a whole reported to the Office of Advanced Research and Technology (OART). The result was that Ames scientists in the life and planetary sciences had little to gain by participating directly in those project efforts, and thus did not compete very hard to get their experiments on either the Pioneers or the Biosatellites. Project management at Ames remained segregated from the laboratory culture of the Center even as it gradually absorbed that culture.

Alfred Eggers, in 1958, at the 10 by 14 inch supersonic wind tunnel.

Transition into NASA: 1959 – 1968

HARVEY ALLEN AS DIRECTOR

On 15 October 1965, DeFrance retired after 45 years of public service, with elaborate ceremonies in Washington and in San Jose so his many friends could thank him for all he had done. DeFrance had planned well for his retirement and had cultivated several younger men on his staff to step into his role. Harvey Allen was the best known of the Ames staff, and had the most management experience. The director's job was his to refuse which, initially, he did.

Eggers then loomed as the front runner. Eggers and Allen were both friends and competitors. Whereas Allen was seen as jovial and encouraging, Eggers was seen as abrasive and challenging. The two had collaborated in the early 1950s on the pathbreaking work on the blunt body concept, but Allen made his work more theoretical whereas Eggers explored practical applications like the lifting bodies. In January 1963, Eggers won for himself the newly created post of assistant director for research and development analysis and planning, where he could pursue his expertise in mission planning. A year later he went to headquarters as

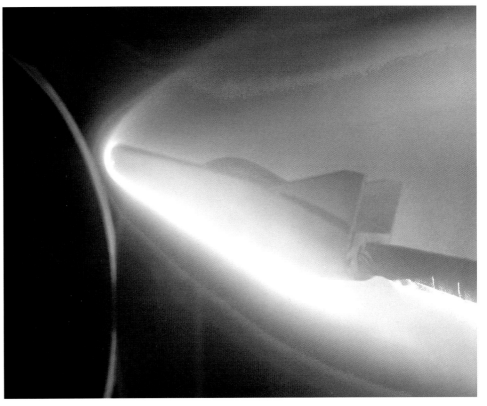

Model of the M-2 lifting body, in 1962, being tested in Ames' atmospheric entry simulator to determine the areas of most intense heat.

deputy associate administrator in OART. He persuaded his boss, Ray Bisplinghoff, to create an OART-dedicated mission analysis group based at Ames. It would report directly to headquarters, be located at Ames, and staffed by scientists on loan from all NASA Centers. But this OART mission analysis division, established in January 1965, never got support from the other Centers. Each Center thought it should bear responsibility for planning the best use of its research and resources. Within a year, the OART abandoned plans for assigning a complement of fifty scientists to the Ames-based OART mission analysis division. But the disarray began to spread to the Ames directorate for R&D planning and analysis that was originally created for Eggers. Clarence Syvertson

Schlieren image of the X-20 Dyna-Soar.

remained in charge of a much smaller, though very active, mission analysis division. A new programs and resources office was created under Merrill Mead to plan and fight for Ames' budget, which left Eggers as the headquarters choice to become director. To prevent that from happening and to keep Ames as it was—distant from Washington, with a nurturing and collaborative spirit, and focused on research rather than projects—in October 1965 Allen took the directorship himself.

Allen did not especially distinguish himself as director as he had in his other promotions. As a person, Allen differed dramatically from DeFrance. He was warm, benevolent, close to the research, inspirational in his actions and words. But Allen, like DeFrance, kept Ames as a research organization and worked hard to insulate his staff from the daily false urgencies of Washington. Allen asked Jack Parsons, who remained as associate director, to handle much of the internal administration and asked Loren Bright and John Boyd to fill the newly created positions of executive assistant to the director and research assistant to the director. Allen often sent Ames' ambitious young stars in his place to the countless meetings at headquarters. And every afternoon at two o'clock, when headquarters staff on Washington time left their telephones for the day, Allen would

H. Julian Allen, Director of Ames Research Center from 1965 to 1969.

Basic design of Pioneer spacecraft 6 through 9.

leave his director's office and wander around Ames. He would poke his head into people's offices and gently inquire about what was puzzling them. "Are you winning?" he would ask.[1] Eventually he would settle into his old office and continue his research into hypersonics.

Ames suffered a bit during Allen's four years as director. Ames' personnel peaked in 1965 at just over 2,200 and dropped to just under 2,000 by 1969. Its budget stagnated at about $90 million. For the first time a support contractor was hired to manage wind tunnel operations—in the 12 foot pressurized tunnel—and there was a drop off in transonic testing and aircraft design research. But tunnel usage actually increased to support the Apollo program, and there was dramatic growth in Ames' work in airborne and space sciences, especially from the Pioneer program.

Pioneers 6 to 9

The Pioneers span the entire recent history of Ames, transcending efforts to periodize them neatly. The first Pioneers—the Pioneer 6 to 9 solar observatories—were conceived under DeFrance and executed under Allen. Allen asked the same group to plan Pioneers 10 and 11, and Hans Mark, Allen's successor as director, presided over the execution of the Pioneers as simple, elegant, science-focused and pathbreaking projects. Every subsequent Ames director—upon the occasion of data returned from some encounter

John Wolfe, Richard Silva and Clifford Burrous in September 1962, with a model of the OGO-1 orbiting geophysical observatory and the solar plasma measuring instrument that they built for it.

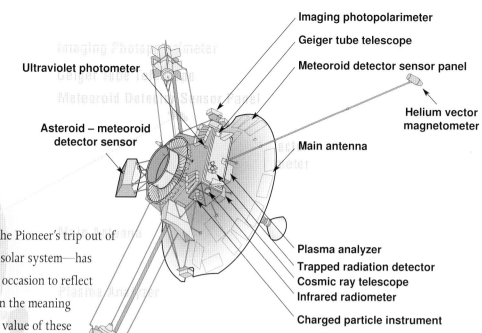

Schematic of Pioneer 10.

on the Pioneer's trip out of our solar system—has had occasion to reflect upon the meaning and value of these sturdy little spacecraft. The Pioneer program is discussed as part of NASA's formative years because, in addition to all the valuable data they produced, in the late 1960s the Ames space projects division devised the Pioneer program as a shot across the bow of the NASA way of doing things.

In 1963, Ames was given a block of four Pioneer flights, and a budget of $40 million to build and launch the spacecraft. The bulk of this funding went to contractors—to Douglas and Aerojet-General to build the Thor-Delta rockets and to Space Technology Laboratories to build the spacecraft. Charlie Hall was the Pioneer project manager at Ames. On 15 December 1965, Pioneer 6 achieved its orbit around the Sun just inside the orbit of Earth. It immediately began sending back data on magnetic fields, cosmic rays, high-energy particles, electron density, electric fields and cosmic dust. It was soon followed by Pioneers 7, 8, and finally Pioneer 9 launched on 8 November 1968.

These four Pioneers sat in different orbits around the Sun, but outside the influence of Earth, and returned data on the solar environment. Until 1972, they were NASA's primary sentinals to warn of the solar storms that disrupt communications and electricity distribution on Earth. When positioned behind the Sun, the Pioneers collected data to predict solar storms since they could track changes on the solar surface two weeks before they were seen on Earth. During the Apollo lunar landings, the Pioneers returned data hourly to mission control, to warn of the intense showers of solar protons which could be dangerous to astronauts on the surface of the Moon.

In addition to building spacecraft and sensors to collect the data, Ames also designed the telemetry to gather the data and the computers to process them. Pioneer 6 first gave accurate measurements of the Sun's corona where the solar winds boil off into space. The plasma wave experiment on the Pioneer 8 provided a

Principal investigators take center stage to explain the results of the Pioneer missions.

full picture of Earth's magnetic tail. For the Pioneer 9 spacecraft, Ames established the convolution coders used for most deep space planetary missions. Since the Sun is typical of many stars, Ames astrophysicists learned much about stellar evolution. Before the Pioneers, the solar wind was thought to be a steady, gentle flow of ionized gases. Instead, the Pioneers found an interplanetary region of great turbulence, with twisted magnetic streams bursting among other solar streams.

As the group that designed and built the early Pioneers then turned their attention to the next space horizon, these simple satellites continued to send back data. Pioneer 9 was the first to expire, in May 1983, well beyond its design lifetime of six months. It had circled the Sun 22 times, in a 297-day orbit. Pioneers 6 and 7 continued to work well into the 1980s, though they were tracked less frequently as newer missions required time on the antennas of NASA's Deep Space Network. By then, these Pioneers had had their days in the Sun.

Pioneers 10 and 11

During the 1960s, astronomers grew excited about the prospects of a grand tour—of sending a space probe to survey the outer planets of the solar system when they would align during the late 1970s. The known hazards to a grand tour—the asteroid belt and the radiation around Jupiter—were extreme. The hazards yet unknown could be worse. So Ames drafted a plan to build NASA a spacecraft to pioneer this trail.

In 1968, the Space Science Board of the National Academy of Sciences endorsed the plan. NASA headquarters funded the project in February 1969, following intensive lobbying by Ames' incoming director, Hans Mark, and Ames' director of development, John Foster. Charles Hall, manager of the Pioneer plasma probe spacecraft, led the project, and asked Joseph Lepetich to manage the experiment packages and Ralph Holtzclaw to design the spacecraft. Chief scientist John Wolfe, who had joined Ames in 1960, did gamma-ray spectroscopy and measurements of the interplanetary solar wind, and later became chief of Ames' space physics branch. Originally called the Pioneer Jupiter-Saturn

Pioneer 10, being tested prior to launch.

A pre-launch view of Pioneer 10 spacecraft, encapsulated and mated with an Atlas-Centaur launch vehicle on 26 February 1972. Pioneers 10 and 11 were ejected from Earth's atmosphere at a greater speed than any previous vehicle.

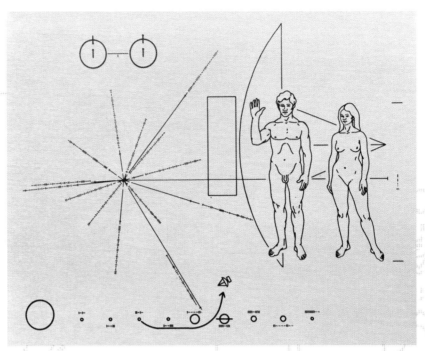

Pioneer 10, the first spacecraft to leave our solar system, carries a message to other worlds. The plaque was designed by Carl Sagan and Frank Drake. The artwork was prepared by Linda Salzman Sagan.

program, upon successful launch the name was changed to Pioneers 10 and 11.

Spacecraft able to explore the giants of our solar system—Jupiter and Saturn—had to be much different from the many spacecraft that had already explored Mars and Venus. First, Jupiter is 400 million miles away at its closest approach to Earth, whereas Mars is only 50 million miles away. Thus, the spacecraft had to be more reliable for the longer trip. Second, since solar panels could not produce enough energy, the spacecraft needed an internal power supply. Finally, the greater distance demanded a larger, dish-shaped high gain antenna.

Added to these more natural design constraints were two early engineering decisions Hall made to keep the project within its budget. Both derived from Ames' experience with the earlier Pioneer plasma probes. First, rather than being stabilized on three axes by rockets, Pioneers 10 and 11 were spin-stabilized by rotating about their axes. The spin axis was in the plane of the ecliptic, so the nine foot diameter communications dish antenna always pointed toward Earth.

Pioneer 10 at TRW in the final stages of assembly.

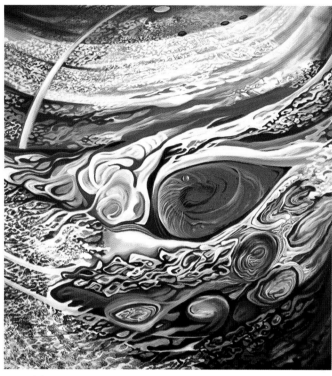

Oil painting depicting the storms of Jupiter, the satellite Io, and the Great Red Spot.

Inertia came from the four heavy nuclear power units—RTGs or radioisotope thermoelectric generators—mounted fifteen feet from the axis on two long beams. Spin stabilization was cheap and reliable, but made high resolution photographs impossible.

The second engineering decision Hall made was to send all data back to Earth in real time at a relatively slow stream of one kilobit per second. Storing data on board was expensive and heavy. This again lowered the resolution of the photographs and the precision of some measurements. It also meant that Pioneer would have to be flown from the ground. Onboard memory could store only five commands, of 22 bits each, needed for very precise maneuvers such as those to move the photopolarimeter telescope quickly during the planetary encounter. Each command had to be carefully planned, since signals from Earth took 46 minutes to reach the spacecraft at Jupiter. Hall convinced the scientists designing Pioneer payloads to accept these limits. They had much to gain, Hall argued, by getting their payloads there on a reliable platform and getting there first.

Charlie Hall leads the Pioneer project staff through an efficient stand-up meeting prior to the encounter with Jupiter.

Eleven experiment packages were hung on the Pioneers, which measured magnetic fields, solar wind, high energy cosmic rays, cosmic and asteroidal dust, and ultraviolet and infrared radiation. (The two spacecraft were identical except that Pioneer 11 also carried a fluxgate magnetometer like the one carried on Apollo 12.) Each spacecraft weighed just 570 pounds, and the entire spacecraft consumed less power

Transition into NASA: 1959 – 1968

Jupiters Red Spot and a shadow of the moon Io, as seen from Pioneer 10.

than a 100 watt light bulb. One of the most significant engineering achievements was in electromagnetic control—the spacecraft was made entirely free of magnetic fields to allow greater sensitivity in planetary measurements.

Ames indeed kept the Pioneers within a very tight budget and schedule. The entire program for the two Pioneer 10 and 11 spacecraft, excluding launch costs, cost no more than $100 million in 1970 dollars. (That compares with $1 billion for the Viking at about the same time.) To build the spacecraft, Ames hired TRW Systems Group of Redondo Beach, California, the company that built the earlier Pioneers. TRW named Bernard O'Brien as its program manager. Hall devised a clear set of management guidelines. First, mission objectives would be clear, simple, scientific and unchangeable. The Pioneers would explore the hazards of the asteroid belt and the environment of Jupiter, and no other plans could interfere with those

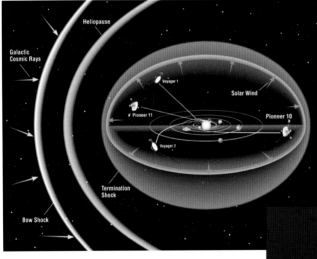

Trajectories of Pioneer 10, Pioneer 11 and Voyager.

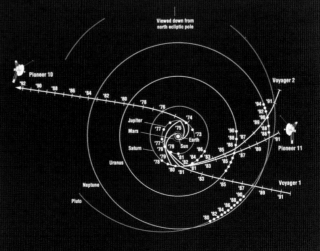

Pioneer 10 encounter with Jupiter.

Jack Dyer and Richard Fimmel in the Pioneer mission control center in May 1983.

goals. Second, the prime contractor was delegated broad technical authority. Third, existing technology would be used as much as possible. Fourth, the management team at Ames could comprise no more than twenty people. Fifth, their job was to prevent escalation of requirements.

One other decision ensured that the Pioneers would have an extraordinary scientific impact. In the 1960s, NASA scientists began to explore ways of flying by gravitational fields to alter spacecraft trajectories or give them an energy boost. Gravitational boost was proved out on the Mariner 10, which flew around Venus on its way to Mercury. Ames proposed two equally bold maneuvers. Pioneer 10 would fly by Jupiter so that it was accelerated on its way out of the solar system, to reconnoiter as far as possible into deep space. Pioneer 11 would fly by Jupiter to alter its trajectory toward an encounter with Saturn five years later. Without diminishing their encounter with Jupiter, the Pioneers could return better scientific data and years earlier than Voyager for the small cost of keeping open the mission control room. No good idea goes unchallenged, and Mark and Hall found themselves lobbying NASA headquarters to fend off JPL's insistence that their Voyager spacecraft achieve these space firsts.

Three months before project launch, Mark got a call from Carl Sagan, the astronomer at Cornell University, a friend of Mark's from time spent at the University of California at Berkeley, and close follower of efforts at Ames to discover other life in the universe. Sagan called to make sure that Mark appreciated "the

PIONEER ★ JUPITER

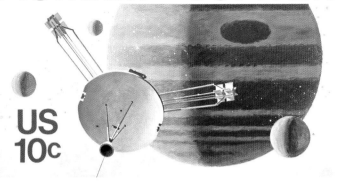

cosmic significance of sending the first human-made object out of our solar system."[2] Sagan wanted the Pioneer spacecraft to carry a message, in case they were ever found, that described who built the Pioneers and where they were from. So Sagan and his wife, Linda, designed a gold-anodized aluminum plate on which was inscribed an interstellar cave painting with graphic depictions of a man, a woman, and the location of Earth in our solar system.

Thirty months after project approval, on 2 March 1972, NASA launched Pioneer 10. Since the spacecraft needed the highest velocity ever given a human-made object—32,000 miles per hour—a solid-propellant third stage was added atop the Atlas Centaur rocket. Pioneer 10 passed the orbit of the Moon eleven hours after liftoff; it took the Apollo spacecraft three days to travel that distance. A small group of five specialists staffed the Ames Pioneer mission operations center around the clock, monitoring activity reported back through the huge and highly sensitive antennas of NASA's Deep Space Network. Very quickly, Pioneer 10 started returning significant data, starting with images of the zodiacal light. On 15 July 1972, Pioneer 10 first encountered the asteroid belt. Most likely the scattered debris of a planet that once sat in that orbit between Mars and Jupiter, the asteroid belt contains hundreds of thousands of rocky fragments ranging in size from a few miles in diameter to microscopic size. From Earth, it was impossible to know how dense this belt would be. An asteroid/meteoroid detector showed that the debris was less

Pioneer 11 pre-encounter with Saturn, as painted by Wilson Hurley.

Artist concept of Pioneer 11 as it encounters Saturn and its rings.

dangerous than feared. Next, in August 1972, a series of huge solar flares gave Ames scientists the opportunity to calibrate data from both Pioneer 10, now deep in the asteroid belt, and the earlier Pioneers in orbit around the Sun. The results helped explain the complex interactions between the solar winds and interplanetary magnetic fields. Ames prepared Pioneer 11 for launch on 5 April 1973, when Earth and Jupiter were again in the best relative positions.

Pioneer 10 flew by Jupiter nineteen months after launch, on 4 December 1973. Over 16,000 commands were meticulously executed on a tight encounter schedule. The most intriguing results concerned the nature of the strong magnetic field around Jupiter, which traps charged particles and thus creates intense radiation fields. Pioneer 10 created a thermal map of Jupiter, and probed the chemical composition of Jupiter's outer atmosphere. Its trajectory flew it behind the satellite Io and, by observing the alteration of the telemetry signal carrier wave, Pioneer 10 provided direct evidence of the very tenuous atmosphere around Io. Signals from the imaging photopolarimeter were converted into video images in real time, winning the Pioneer project an Emmy award for contributions to television. Most important, Pioneer 10 proved that a spacecraft could fly close enough to Jupiter to get a slingshot trajectory without being damaged.

Pioneer 11 flew by Jupiter a year after Pioneer 10. In November 1974, its encounter brought it three times closer to the giant gas ball than Pioneer 10. Ames mission directors successfully attempted a somewhat riskier approach, a clockwise trajectory by the south polar region and then straight back up through the intense inner radiation belt by the equator and back out over Jupiter's north pole. Thus, Pioneer 11 sent back the first polar images of the planet. Pioneer 11 reached its closest point with Jupiter on December 3, coming

John Wolfe describes the transit of Pioneer 10 around Jupiter.

within 26,000 miles of the surface. This mission gathered even better data on the planet's magnetic field, measured distributions of high-energy electrons and protons in the radiation belts, measured planetary geophysical characteristics, and studied the Jovian gravity and atmosphere. Pioneer 11 then continued on to its encounter with Saturn on 1 September 1979. There it discovered a new ring and new satellites, took spectacular pictures of the rings around Saturn, and returned plenty of data about Saturn's mass and geological structure.

Pioneer 10, meanwhile, continued on its journey out of the solar system. On 13 June 1983 it passed the orbit of Pluto. The Pioneer project team, now led by Richard Fimmel, eagerly looked for any motion in its spin stabilized platform that would indicate the gravitational pull of a tenth planet, but found none. On its 25th anniversary in 1997, Pioneer 10 was six billion miles from Earth, still the most distant of human-made objects, and still returning good scientific data. By 1998, it had still not detected the plasma discontinuity that defines the edge of the heliopause, where the solar winds stop and our Sun no longer exerts any force. Pioneer was so far from Earth that its eight watt radio signal, equivalent to the power of a night light, took nine hours to reach Earth. The closest approach to any star will be in about 30,000 years, as Pioneer flies by the red dwarf star Ross 248.

A global mosaic of Saturn during the Pioneer 11 encounter. The irregular edge of the ring is caused by stepping anomalies of the imaging photopolarimeter.

The engineering model for the Pioneers hangs in the Hall of Firsts at the National Air and Space Museum since the actual Pioneers were, in fact, the first human-made objects to leave our solar system. They are also honored as the spacecraft that paved the way for exploration beyond Mars. NASA eventually did fund the grand tour, with spacecraft much different from the Pioneers. Voyagers I and II, designed and managed at JPL, were sophisticated and stable platforms that weighed more than 2,000 pounds, cost $600 million to develop, and carried better cameras to return more spectacular photographs. Ames people will always remember the Pioneers, by contrast, as spacecraft that flew much the same mission, but faster, better, and cheaper. These spacecraft—simple in concept, elegant in design, competently executed and able to return so much for so little—served as models for the spirit Ames would infuse into all of its work.

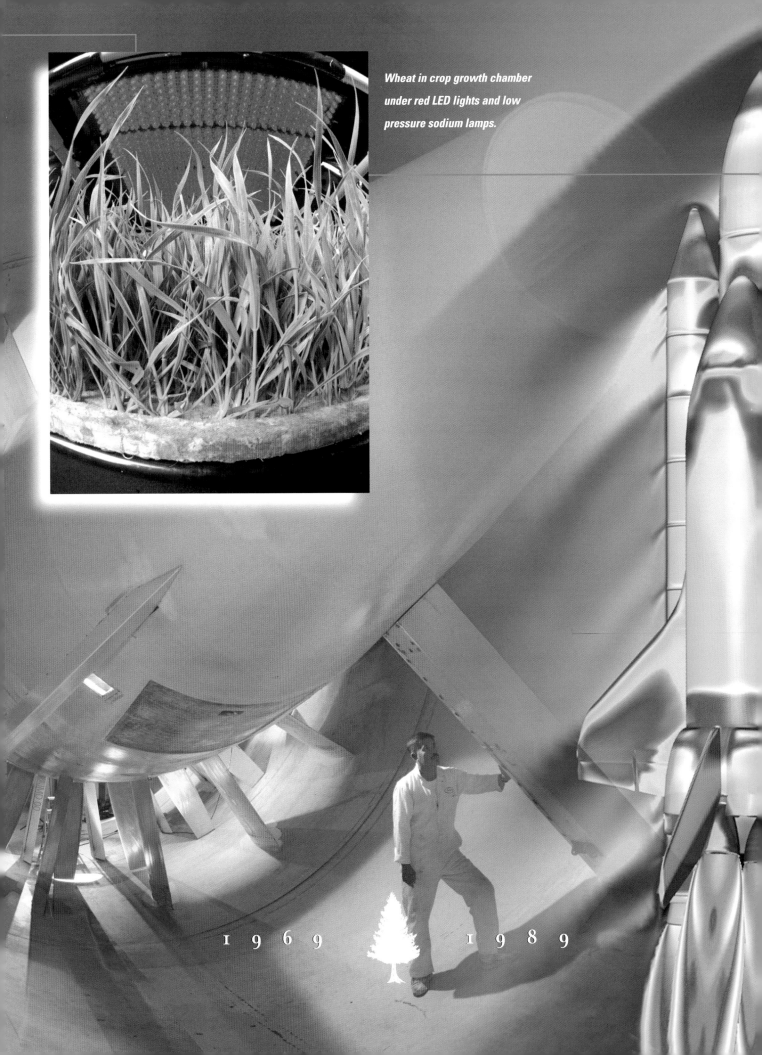

Wheat in crop growth chamber under red LED lights and low pressure sodium lamps.

1 9 6 9   1 9 8 9

DIVERSE CHALLENGES EXPLORED WITH UNIFIED SPIRIT

Chapter 3:
Ames in the 1970s and 1980s

Michael McGreevy holding a televised rock in his virtual hand using the Ames EXOS Dexterous Interface in May 1992.

Two events make 1969 the year to mark the next era in Ames history. First, Apollo 11 returned safely from its landing on the Moon, signalling the beginning of the end of the lunar landing mission that drove NASA almost from its start. NASA had yet to decide what to do for its second act, and its flurry of strategic planning took place against an uncertain political backdrop. Much of the American public—including both political conservatives concerned with rampant inflation and political liberals concerned with technocratic government—began to doubt the value of NASA's big plans. NASA had downplayed the excitement of interplanetary exploration as it focused on the Moon. Congress and the American aerospace industry, under pressure from a resurgent European aerospace industry, began to doubt if NASA really wanted the aeronautics part of its name. NASA had to justify its budget with quicker results, better science, and relevance to earthly problems.

The second major event of 1969 was the arrival of Hans Mark as Ames' director. Mark, himself, displayed a force of personality, a breadth of intellect, and an aggressive management style. More significantly, Mark arrived as rumors circulated that Ames would be shut down. Thus, Ames people gave him a good amount of room to reshape their institution. An outsider to both Ames and NASA, Mark forged a vision for Ames that nicely translated the expertise and ambitions of Ames people with the emerging shape of the post-Apollo NASA. Mark fashioned Ames to epitomize the best of what NASA called its OAST Centers—those reporting to the Office of Aeronautics and Space Technology. Mark left Ames in August 1977, but became in effect an ambassador for the Ames approach to research management during his posts at the Defense Department and at NASA headquarters.

Into the 1970s, NASA increasingly focused its work on the Space Shuttle, taking the posture that access to space would soon be routine. Ames responded to this mission, first, by creating technologies that would make the Shuttle as routine as other aircraft and second, by showing that there was still room within NASA for the extraordinary. This was a period at Ames when what mattered most was entrepreneurship, reinvention and alliance-building. Ames reshaped itself, so that the institutional structures that mattered most included the Ames Basic Research Council, the Ames strategy and tactics committee,

Hans Mark, Director of Ames Research Center from 1969 to 1977.

quality circles, and consortia agreements between Ames and universities. Ames more consciously developed its people, so that Ames people played ever larger roles in NASA administration.

REDEFINING THE DIRECTORS' ROLE: HANS MARK

Like Ames directors tend to be, Hans Mark was a practicing researcher. But he was the first senior executive at Ames who did not come up through its ranks. Mark was born 17 June 1929 in Mannheim, Germany, and emigrated to America while still a boy. He got an A.B. in 1951 in physics from the University of California and a Ph.D. in 1954 in physics from the Massachusetts Institute of Technology (MIT). He then returned to Berkeley and, save for a brief visit to MIT, stayed within the University of California system until 1969. He started as a research physicist at the Lawrence Radiation Laboratory in Livermore and rose to lead its experimental physics division. He also rose through the faculty ranks at the Berkeley campus to become professor of nuclear engineering. In 1964 he left his administrative duties at Lawrence Livermore National Laboratory to become chair of Berkeley's nuclear engineering department as it shifted its emphasis from weapons to civil reactors.

When he arrived at Ames, Mark applied many of the management techniques he had witnessed at work in the nuclear field. He created a strategy and tactics committee that allowed for regular discussions among a much broader group than just upper management, about where Ames was going and what would help it get there. As a result, Ames people became very good at selecting areas in which to work. Tilt rotors, for example, brought together a wide array of research at Ames to tackle the problem of air traffic congestion. Ames deliberately pioneered the new discipline of computational fluid dynamics by acquiring massively parallel supercomputers and by merging scattered code-writing efforts into a coherent discipline that benefitted every area at Ames.

Similarly, Mark created the Ames "murder" board. This board was a sitting group of critics that questioned anyone proposing a new project or research area, to toughen them up for the presentations they would make at headquarters. His style was argumentative, which he thought Ames needed in its cultural mix. In a period of downsizing, Mark wanted Ames people to stake out "unassailable positions"—program areas that were not just technically valuable but that they could defend from any attack.

From his experience at Livermore, Mark also understood the power of matrix organization, the idea then underlying the restructuring of all research and development in the military and high-technology industry. Though formal matrix organization fitted Ames badly—because of the traditional structure around disciplinary branches and functional divisions—Mark used the strategy and tactics committees to get people thinking about the on-going relationship between functional expertise and time-limited projects. Ames took project management more seriously, using the latest network scheduling techniques to complement its tradition of foreman-like engineers. And Ames bolstered the functional side of its matrix, by getting its scientific and facilities staffs to more consciously express their areas of expertise.

Ames people insisted that Mark understand that they were each unique—willing to be herded but never managed. Mark compromised by mentally grouping them as two types. Some wanted to become as narrow as possible in a crucial specialty that only NASA would support, because academia or industry would not. Mark admired these specialists, but took the paternal attitude that they were incapable of protecting themselves. The other type warmed to the constant and unpredictable challenges of aerospace exploration and constantly reinvented themselves. So Mark created an environment of opportunities, perhaps unique in NASA, where both types of researchers flourished. And Mark

The Director's staff was a training ground for future NASA leaders. At Mark's farewell party in 1977 are (left to right) Alan Chambers, Dale Compton, Jack Boyd, Hans Mark, Lloyd Jones and John Dusterberry.

adopted the Ames custom of motivation and management by meandering. Like Harvey Allen before him, Mark poked his head randomly into offices to ask people what they were up to, and took it as his responsibility to understand what they were talking about. When he did not have time to stride rapidly across the Center, he would dash off a handwritten memo (that people called Hans-o-grams) that concisely presented his point of view. When scientists like R. T. Jones and Dean Chapman suggested that Mark could know a bit more about the work done at the Center, they convened a literature review group that met every Saturday morning after the bustle of the week. While at Ames Mark learned to fly just so he could argue with aerodynamicists and flight mechanics.

Mark treated NASA headquarters in the same informal way. He encouraged Ames people to see headquarters as more than an anonymous source of funds and headaches. Mark showed up every morning at six o'clock so his workday was synchronized with eastern time. He travelled constantly to

R. T. Jones (right) in February 1975 preparing a model of his oblique wing aircraft for tests in the Unitary plan tunnels.

Washington D.C., taking a red eye flight there and an evening flight back. He attended every meeting he thought important and told anyone who would listen how Ames was shaping its future. There, too, he would poke his head randomly into offices to chat with the occupants about how to shape NASA strategy. To head the Ames directorates of aeronautics, astronautics, and life sciences, Mark picked entrepreneurs who were likewise willing to travel and sell. Their deputies would stay home and manage daily operations. From Mark, headquarters got the impression that Ames would be more involved in deciding how its expertise would be used. They also got the impression that Mark had a "stop me if you can" attitude toward headquarters and shared little respect for chains of command.

Mark also made Ames collaborate with broader communities. NASA headquarters was often too rule-bound or unimaginative to fund every program Ames wanted to accomplish. Collaboration increased the opportunities for direct funding. Collaboration also made Ames people think about the larger scientific and educational constituencies they served, and increased the chances that all the best people would contribute to Ames' efforts. Mark broke open the fortress mentality that DeFrance had inculcated, and encouraged everyone to put out tentacles in whatever direction they thought appropriate.

During Mark's tenure Ames forged on-going ties with universities. While Ames had long used individual contracts with area universities for specific types of help, in 1969 Ames signed a cooperative agreement with Santa Clara University that was open-ended. Negotiated earlier by Ames chief counsel Jack Glazer, it further pushed the limits of the Space Act of 1958.

The Ames spirit of free and vigorous discussion. Left to right: R.T. Jones, Jack Nielsen, Hans Mark, Leonard Roberts and Harvey Allen.

Atmosphere of Freedom Sixty Years at the NASA Ames Research Center

Hans Mark with Edie Watson.

The agreement defined an on-going infrastructure of collaboration so that Ames and university scientists only needed to address the technical aspects of their work together. Furthermore, students could come to Ames to write their dissertations, and many did in the fields of lunar sample analysis and computational fluid dynamics. Some students came to write papers on the law of space, or intellectual property, since Glazer had made his office the only legal counsel office in NASA with a research budget. Rather than getting a contract with research bought solely for NASA's benefit, the collaborating universities shared substantially in the cost of research. Ames signed collaborative agreements with universities around America so that in June 1970, when President Nixon tried to appoint a government czar of science to keep university faculty out of the pockets of mission-oriented agencies like NASA, Ames stood out as exemplary on the value of collaboration at the local level. In 1971, headquarters let Ames award grants as well as administer them; by 1976, Ames' university affairs office could administer the grants independent of the procurement office. By 1978, Ames administered 260 grants to 110 universities with annual obligations of more than $11 million.

Mark also encouraged Ames researchers to interact more freely with engineers in industry, and allowed them more freedom to contract with the firms most willing to help build products for Ames' needs. Mark encouraged the Army to augment its rotorcraft research office at Moffett Field, and opened dialogue with the Federal Aviation Administration (FAA) about joint programs. Mark put the Illiac IV supercomputer on the Arpanet to encourage a much wider community to

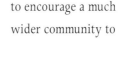

Ames has long studied ways to improve the impact of air traffic on communities. In 1974, Ames researchers set up a test of noise patterns propagating from a model of a transport aircraft.

Aerospace Encounter at Ames Research Center became a reality in October 1991.

write its code. And he encouraged Ames senior staff to seek advancement throughout the Administration. He was especially proud that people nurtured in Ames' atmosphere were named director at Lewis and Goddard (John Klineberg), director at Langley (Richard Peterson), associate administrator for management at headquarters and deputy director at Dryden (John Boyd).

Mark left Ames in August 1977, having guided Ames people to shape a long-term vision of where they wanted to go. He helped match their creative energy with NASA's larger and ever-shifting ambitions. The next three directors of Ames shaped the Center in the same way, but with an ever evolving palette of personnel against a changing canvas of scientific progress and international politics. Although none hit Ames with the same amount of cultural dissonance, each of these directors learned his approach by watching Mark at close range. In fact, Mark's very first decision as director was to confirm the decision by NASA headquarters that his deputy would be Clarence Syvertson.

Clarence A. Syvertson

Clarence "Sy" Syvertson understood the NACA culture that had made Ames so great. He arrived at Ames in 1948, after taking degrees at the University of Minnesota and after a stint in the Army Air Forces, to work with Harvey Allen in solving the problems of hypersonic flight. Syvertson then worked with Al Eggers in the 10 by 14 inch wind tunnel until 1959, when he was named chief of the 3.5 foot hypersonic tunnel that he designed. By pioneering theories that could be tested in Ames' complex of wind tunnels, Syvertson outlined the aerodynamic limits for some aircraft that NASA still hopes to build—a hypersonic skip glider, direct flight-to-orbit aircraft, and hypersonic transports. For the North American B-70 bomber, he defined the high-lift configuration later incorporated in other supersonic transport designs. Syvertson also managed the design and construction of the first lifting body, the M2-F2, a prototype wingless aircraft that could fly back from orbit and land at airfields on Earth. A successful series of flight tests in 1964 with the M2-F2 guided the configuration of the Space Shuttle orbiter.

In 1964 Syvertson created and led the NASA mission analysis division, based at

Diverse Challenges Explored with Unified Spirit: 1969 – 1989

Clarence A. "Sy" Syvertson, Director of Ames Research Center from 1977 to 1984.

Ames but staffed by all of the NASA Centers, which charted dramatic new ways to explore the outer planets. In 1966 Syvertson became director of astronautics, then in 1969 became deputy director of Ames. Syvertson was awarded NASA's Exceptional Service Medal in 1971 for serving as executive director of a joint DOT-NASA policy study that made key recommendations on civil aviation and helped move Ames into air traffic issues.

As Mark's deputy, Syvertson was the inside man. He managed the internal reconfiguration of Ames so that Mark could focus on its future and on its relations with Washington. He managed renovation of the main auditorium so that Ames people could present lectures, award ceremonies and media events in a better setting. Syvertson was known as a consensus-builder—able to step in, forge compromise, and resolve the conflict that Mark had encouraged, be it policy warfare with headquarters or argumentation internally. When Mark left Ames in 1977, NASA headquarters actually advertised the job of Ames director. After a year, the "acting" was removed from Syvertson's position and he was made permanent director because, many people noted, Ames could not survive another Mark.

Ames grew more slowly during Syvertson's tenure, and the pace of contracting in support services accelerated. But Syvertson broke ground for some important new facilities at Ames—like the crew-vehicle systems research facility and the numerical aerospace simulation facility—and extended its collaboration in new areas. Syvertson accelerated Ames' outreach efforts, especially to pre-college students. The teacher resource center, for example, archived slides, videos and other media that science educators could borrow to improve their classes. Class tours grew more frequent, so Syvertson

Grumman F-14A model undergoing tests in 1970 in the 9 by 7 foot wind tunnel.

helped form a hands-on teaching museum, which ultimately opened in October 1991 as the Ames Aerospace Encounter built in the old 6 by 6 foot wind tunnel.

Perhaps the biggest challenge to Syvertson and Ames management came in 1981 with Ames' consolidation of the Dryden Flight Research Center. Soon after headquarters had sent Ames' aircraft to Rogers Dry Lake in 1959, Ames started adding aircraft back to its fleet at Moffett Field—first helicopters and vertical takeoff and landing aircraft, then airborne science platforms. When the Reagan administration demanded that NASA cut its staff by 850, acting administrator A. M. Lovelace responded by implementing a plan to make Wallops Island Flight Center an administrative unit of Goddard and to make Dryden an "operational element and component installation" of Ames.[1] The merger, effective 1 October 1981, formalized a strong relationship. Ames researchers already performed most of their test flights at Dryden; and most of Dryden's flight projects originated at Ames. Both of the Ames-based tilt rotors had been flying at Dryden, and Ames willingly transferred more research aircraft there now that its staff was ultimately in charge.

The consolidation was implemented by Louis Brennwald, as Ames' director of administration, with consolidation planning led by John Boyd, then Ames' associate director and a deputy director at Dryden from 1979 to 1980. Both aeronautics and flight systems directorates were completely reorganized, without requiring reductions in force or involuntary transfers. Consolidation meant that Dryden administered flight operations there, where it was cheaper and safer, and Ames provided technical leadership and policy guidance.

The Ames–Dryden Flight Research Facility sat on the edge of Rogers Dry Lake, a vast, hard-packed lake bed near the town of Muroc in the Mojave desert of southern California. Its remote location, extraordinarily good flying weather, exceptional visibility and 65 square mile landing area all made it a superb test site. Edwards Air Force Base managed the site, and NASA's Western Aeronautical Test Range provided the tracking and telemetry systems to support the research. Ames–Dryden also

Teachers tour the Ames plant growth facility.

Diverse Challenges Explored with Unified Spirit: 1969 – 1989

The Ames-Dryden Flight Research Facility.

ran the world's best facility for remotely piloted flight, and its flight loads research facility allowed ground-based structural and thermal tests of aircraft, as well as calibration of test equipment. With better access to Dryden facilities, Ames researchers more efficiently moved innovative designs from concept to flight. To move from concept to flight, Ames had computational power for aerodynamic design and optimization, wind tunnels for measuring loads and fine-tuning configurations, simulators to study handling qualities, and shops to build the proof-of-concept vehicles. The best examples of Ames' abilities to move ideas into flight quickly and cheaply are the AD-1 oblique wing aircraft, the HiMAT remotely piloted high g research vehicle, and the F-8 digital fly-by-wire program.

Eventually, Ames had to address the Reagan administration's demand for staff cuts. In 1983 a program review committee led by deputy director Gus Guastaferro decided to cut back on new space projects to support existing ones, and to mothball several research facilities—the 14 foot tunnel, the 3.5 foot hypersonic tunnel, the transportation cab simulator, and the vertical acceleration and roll device. Yet Ames continued to pursue the same broad areas it had staked out as unassailable in the early 1970s. Aeronautical research focused on testing methodologies, safety studies, and slow-speed technologies and vertical takeoff aircraft. Space research focused on thermal protection and spacecraft configurations, adding infrared astronomy and airborne sciences, as well as extending the Pioneer efforts into probes of

AD-1 oblique wing aircraft in flight over the Ames-Dryden Flight Research Facility.

planetary atmospheres. All Ames research efforts were infused with its ability to build outstanding laboratory tools—wind tunnels, test models, and motion and work simulators. Most notably, supercomputing permeated everything so that computer codes seemed to replace the scientific theory that had earlier guided so much of what Ames did. By Syvertson's retirement in January 1984, Ames had bolstered its prominence within NASA and among wider communities.

FLIGHT RESEARCH

Leonard Roberts served as Ames' director of aeronautics and flight systems from 1972 through 1984, when integrative projects dominated. He helped match all Ames facilities—the tunnels, computers, simulators and the test grounds at Dryden—with important new flight research programs in maneuverability, short takeoff and landing aircraft and aircraft safety. Ames grew especially adept at building light, inexpensive and well-focused flying laboratories to verify component technology, to test seemingly bizarre new configurations, or to gather data that could not be gathered otherwise.

Another airborne research platform arrived at Ames in April 1977. Lockheed originally built the YO-3A as an ultra-quiet spy plane. The sailplane wings, muffled engine, and slow-turning, belt-driven propeller kept the YO-3A quiet enough that Ames and Army

Oblique wing model mounted in the 11 foot wind tunnel with R. T. Jones. The asymmetrical design allows the aircraft to fly faster, yet consume less fuel and generate less noise than traditionally winged aircraft.

Diverse Challenges Explored with Unified Spirit: 1969 – 1989

SH-3G helicopter and YO-3A observation aircraft in flight above the Crows Landing auxiliary airfield during tests of an airborne laser positioning system.

researchers could add microphones to the wing-tips and tail-fin to accurately measure noise from nearby aircraft. Ames and Army researchers used the converted YO-3A primarily for studying helicopter noise. The test aircraft flew behind the YO-3A, while onboard aero-acoustic measurements were synchronized with data on flight and engine performance telemetered from the test aircraft. Again, based on this research, the FAA asked Ames to play a larger role in research in minimizing flight noise.

Digital Flight Controls

Exemplifying this integrative urge—especially between researchers at Ames and Dryden—was digital fly-by-wire technology (DFBW). Engineers at Ames had pioneered the concept of digital fly-by-wire in the 1960s, expecting it to replace the heavy and vulnerable hydraulic actuators still used in high-performance aircraft. Ames had already designed many of the electronic controls for its ground-based flight simulators, which were made digital in order to run programs stored on their computers. Making these codes and controls reliable enough for a flying aircraft, however, required a magnitude greater of integration and testing. So they acquired an F-8 Crusader fighter aircraft, removed all the mechanical controls, and installed their best DFBW technology. In 1972 in the air above Ames–Dryden, they first demonstrated the system.

Once Ames had demonstrated the feasibility of DFBW, they worked to provide hardware and programming code that the aerospace industry could use in any new aircraft. Any bug in a multiple channel digital system, like DFBW, could crash all redundant channels and all redundant hardware. To avoid the cost, complexity and weight of a backup system, Ames designed software that could survive any problem in the main program. They further designed this fault-tolerant software to check itself automatically

NASA Learjet in flight.

during flight. Other Ames specialists in computational fluid dynamics applied algorithms that incorporated nonlinear functions into the software, and thus allowed DFBW to expand the flight envelope to the extremes of turbulence and boundary layer separation. The success of fly-by-wire in the Ames–Dryden tests convinced NASA to use it as the Space Shuttle flight control system.

Ames next applied its skills to the equally complex, multichannel task of controlling jet engines. Ames designed a digital electronic engine control (DEEC) that could optimize the ten variables on the F100 engine that powers the F-15 and F-16 fighter aircraft. Electronic control greatly improved engine performance, with higher thrust, faster throttle response, improved afterburner response, stall-free operation, and eight times better reliability and maintainability than the mechanical controls it replaced. The Ames DEEC first flew on a NASA F-15 in 1982 and, suitably revised by McDonnell Douglas to military specifications, entered production on U.S. Air Force models in 1985.

Ames continued integrating the components of its digital control technology. In the skies over Ames–Dryden, on 25 June 1985, a group of engineers led by program manager Gary Trippensee witnessed the first flight of the NASA F-15 which had been modified as the HIDEC aircraft (for highly integrated digital electronics control). By integrating data on altitude, Mach number, angle of attack and sideslip, HIDEC let the aircraft and engine operate very close to the stall boundary. Simply by improving these controls and reducing the stall margin, thrust improved two to ten percent. The next phase of the Ames–Dryden HIDEC program included flightpath management, by adding a digital flight controller built by Lear Siegler Corp. for the Air Force. This technology optimized trajectories to minimize fuel consumption, suggest faster intercepts, and allow navigation in four dimensions. The FAA asked Ames to expand upon this flightpath controller in order to help improve capacity in the commercial airspace system. So Ames developed a set of algorithms to process data from aircraft sensors into cockpit instructions on how a pilot could fly more efficiently. The Ames algorithm found its way onto the new Boeing 757 and 767 and Airbus A310 aircraft, and the airlines estimated that Ames' work saved them four percent on fuel costs.

Ames–Dryden staff then used the F-15 flight research aircraft to develop self-repairing flight controls. In flight tests during May 1989, sensors and computers aboard the F-15 correctly identified a simulated failure in the flight controls. Diagnosing failures on the ground is always time consuming, and often fruitless since the failures can only be identified during specific flight conditions. Once the system identified the failure, it could reconfigure other parts of the aircraft to compensate.

Ames powerful triad of facilities—tunnels, computers and simulators—allowed it to create and prove the fundamental hardware and software that controls all recent aircraft. It created protocols useful in the increasing integration of electronics and software in flight systems. And it validated the use of airborne laboratories—like the F-8 and the F-15—to quickly and cheaply validate the importance of component technologies.

Research Aircraft

Ames also drove development of new experimental aircraft. In the early 1960s, for example, Ames aerodynamicist R. T. Jones worked out the theory behind the oblique wing. The wing was perpendicular to the fuselage at takeoff to provide maximum lift, then swiveled in flight so that one half-span angled forward and the other angled backward to decrease drag. This shape could solve the transonic problems of all naval aircraft, which

Three ER-2 aircraft in flight over Ames.

Ames' 1981 fleet of aircraft and helicopters in front of the main hangar.

needed high lift to get off a carrier and a sleek profile to go supersonic. Swept wings, like those on the F-111, solved this problem by using a joint that was heavier and weaker than the swivel joint needed to support an oblique wing. Plus, the oblique wing was extremely efficient in its lift-to-drag ratio at supersonic speeds.

Aerodynamically, however, the oblique wing was very complex. First, the airfoil had to provide lift with air moving over it at a variety of angles. Second, flight controls had to be sophisticated enough to compensate for the asymmetry of the control surfaces. Ames' ongoing work in digital fly-by-wire made it easier to design the oblique wing, by enabling programmers to write code to control an inherently unstable aircraft.

Jones had already established his reputation in theoretical aerodynamics. He saw in the oblique wing not only a promising concept and an intellectual challenge, but also a program to validate Ames' integrative approach to flight research. Jones marshalled the full scientific resources of Ames—especially its wind tunnels and computer modelling—to design the experimental aircraft called the AD-1 (for Ames–Dryden). Then, the AD-1 was fabricated quickly and cheaply, using sailplane technology and a low speed jet engine. By taking this low cost approach, Jones quickly validated the concept and assessed flying qualities without the bureaucratic squabbles that usually accompany X-series aircraft.

Ames' DC-8.

F-18 installed in the 80 by 120 foot test section for tests at high angles of attack, September 1993.

The HiMAT, which first flew in July 1979, was specifically designed for flight tests of high maneuverability concepts. HiMAT (for highly maneuverable aircraft technology test bed) was a Dryden project until Ames was called in to help solve some aerodynamic problems. William Ballhaus wrote the codes to solve three-dimensional, transonic, small perturbation equations that marked the first time that computational fluid dynamics (CFD) had been used to design a wing. (Later this code was used to design the wing for the Sabreliner and for the B-2 stealth bomber, establishing Ballhaus' reputation in applied CFD.) Dryden and Ames staff designed the HiMAT as a small scale, remotely piloted, and heavily instrumented aircraft to test out risky technology. At a fraction of the time and cost of a human-carrying vehicle, Ames tested the interactions between many new high maneuverability devices on an aircraft in flight.

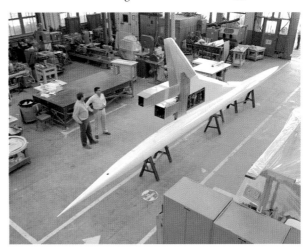

Fabrication in the Ames model shop of a semi-span model for the HEAT project to develop high-lift engine aeroacoustic technology.

HiMAT included digital fly-by-wire, relaxed static stability, close coupled canards and aeroelastic tailoring. Aeroelastic tailoring of composite materials allowed Ames to construct wings so that airflows twisted them to the optimum camber and angle, whether at cruise speeds or undergoing heavy wing loading during maneuvers. Tests of aeroelastic tailoring on the HiMAT provided valuable data on the use of composite materials in all modern aircraft.

Flight Test Technologies

Perhaps because Ames people directed work at Dryden, there was a flourish of research into ways of improving the correspondence between tunnel tests and flight tests. For example, Ames designed a remotely augmented vehicle to expand its skills in flight test instrumentation. This vehicle collected data using the same sensors that collected data

Model being installed in the 40 by 80 foot wind tunnel to test high-lift engine aeroacoustic technology.

during flight tests, telemetered it to a computer on the ground, which transmitted back commands to the flight controls to augment the aircraft's performance. This ground-based computer was easy to maintain and upgrade, flexible enough to control several test aircraft, and powerful enough to run more sophisticated software than was possible on flight-approved computers. Ames used this technology to test new artificial intelligence algorithms before preparing them for inclusion in flight controllers. And it proved a far more efficient way in which to take the next step forward in variable stability flight-test aircraft.

Similarly, Ames' flight test autopilot was a digital computer into which engineers programmed an exact flight maneuver. Since this test autopilot was patched directly into the onboard flight controls, there was no need for additional actuators. The pilot could, of course, override it at any time, but it proved especially valuable when a pilot had to simultaneously perform many maneuvers and control many flight variables, or when repeatability of a maneuver was important.

Ames–Dryden pilots also developed the technology of the transition cone. To scale results from wind tunnel models up to full-scale aircraft, aerodynamicists needed to understand where boundary layers made the transition from laminar to turbulent flow. Researchers at the Arnold Engineering Development Center originated the transition cone concept, which pilots and engineers at Ames–Dryden then tested at a variety of Mach numbers in wind tunnels and mounted to the nose cone of

The NFAC, in November 1984, with the new 80 by 120 foot section added.

Diverse Challenges Explored with Unified Spirit: 1969 – 1989

NASA's F-15. They obtained data that set standards, used worldwide, on the quality of airflows in wind tunnels.

NASA's high-alpha technology program was an effort to calibrate its many research tools while exploring an intriguing regime of aerodynamics. For twelve weeks beginning in June 1991, an Ames team led by Lewis Schiff tested a Navy F/A-18 in the 80 by 120 foot section of the National Full-Scale Aerodynamics Complex (NFAC), making it the first full-scale aircraft tested in the world's largest wind tunnel. The goal was to understand how a modern fighter aircraft performed at very high angles of attack (called high alpha) like those encountered in aerial combat. Wind tunnel data were matched against the data predicted by computational fluid dynamics, and both were compared with flight test data collected on a highly instrumented F/A-18.

FAA/DOT/NASA Safety, Workload, and Training Studies

Beginning with its first research effort, in aircraft de-icing, Ames had pursued specific projects to make aircraft safer and more efficient. Into the 1970s, Ames attacked the problems of aircraft safety with a comprehensive agenda of research projects.

Ames opened its flight simulator for advanced aircraft (FSAA), in June 1969, initially to analyze concepts for the cockpits of the Space Shuttle and fighter

The Ames outdoor aerodynamic research facility (OARF), during 1995 trials of Lockheed's X-32 tri-service lightweight fighter aircraft.

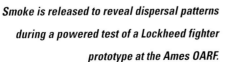

Smoke is released to reveal dispersal patterns during a powered test of a Lockheed fighter prototype at the Ames OARF.

This high Reynolds number channel was opened in 1980 to complement a channel opened in 1973. It is a blow-down facility, with a test section like a wind tunnel, but the flow comes from compressed air on one end shooting into vacuum balls at the other end. The walls of the test section are flexible so they can be adjusted to minimize wall interference with the airflow. Aerodynamicists used it for experimental support—to verify computational fluid dynamics codes and for very precise studies of two-dimensional airflows.

aircraft. It soon became the key part of an increasingly comprehensive collection of facilities dedicated to flight simulation and was used to conduct experiments on how to improve pilot workloads, aircraft automation, flight safety, airline efficiency and, later, air traffic control. Ames researchers then broadened its use to encompass the entirety of the national airspace system. Ames built an alliance with the FAA, which had a research laboratory for its applied research but did little basic science, and with the newly created Department of Transportation which had not yet developed its research capability. Ames brought into this flight safety partnership the full range of its capabilities—in communications, simulation, materials science and computing.

The FAA asked Ames, for example, to devise an aviation safety reporting system (ASRS) to collect data—supplied voluntarily by flight and ground personnel—on aircraft accidents or incidents in U.S. aviation. The Ames human factors group, led by Charles Billings, brought every involved group into the planning, and ASRS director William Reynard implemented it expertly and fairly. The ASRS won the trust of pilots and air traffic controllers, who initially balked at reporting incidents because these incidents almost always arose from simple human error. Ames did not collect the data anonymously, since they had to verify them, but removed identification before compiling data for the FAA. In its first fifteen years, ASRS received 180,000 safety reports, at a rate of 36,000 a year by 1991. From this massive database on human performance in aviation, Ames staff generated hundreds of research papers that led to improvements in aviation safety. The ASRS also put out periodic alert messages about matters that required immediate attention, and a monthly safety report.

Civil transport model installed in the 40 by 80 foot wind tunnel for low speed tests.

"There's nothing worse than sending information to a government agency," said Reynard, "and seeing nothing happen."[2]

Using these data to locate weak spots in the system, Ames used its simulators to minimize human errors. One protocol tested on the simulators became known as line-oriented flight training (LOFT), a method devised at Ames for training crews in all facets of a flight. Previous methods of training and crew testing focused on their response to emergency situations. Because they were maneuver oriented, these methods tended to generate programmed responses. Line-oriented training used a large scale simulator which recreated an entire flight from point to point, interjecting complex problems along the way to test the coordination of decision-making. Airlines adopted a version of it, as did the U.S. Air Force and the FAA.

In the late 1970s, during the flight simulations underlying line-oriented training, Ames discovered that most accidents occurred not because pilots lacked technical skill, but because they failed to coordinate all the resources available in the cockpit. Paradoxically, most training focused on technical proficiency with individual parts of the cockpit. So the Ames aeronautical human factors branch developed methods for use in training pilots to manage all cockpit resources. Ames and the U.S. Air Force Military Airlift Command organized a conference, attended by more than 200 aircraft safety experts from 14 countries, that established the importance of training pilots in cockpit resource management.

Control room for the 40 by 80 foot wind tunnel, June 1992.

This work then led to better workload prediction models, which Ames used to devise simulation scenarios subjecting pilots to standardized workloads. From this, the U.S. Air Force adopted a single code (to simulate supervisory control) to promote its pilots, and NASA adopted

Laser velocimeter, operated by Mike Reinath to visualize flow patterns around models in the 7 by 10 foot wind tunnel, January 1987.

Laser tracking for a model flying through the hypervelocity free flight facility.

Studies to decrease the drag and improve fuel efficiency of a tractor trailer took place in October 1988 in the 80 by 120 foot wind tunnel.

a target-selection code to evaluate control devices for the Space Shuttle. Ames continued to study theories of cockpit automation to reduce pilot error, and then built these into the crew vehicle systems research facility (CVSRF). Opened in 1984, the CVSRF encompassed all facets of air traffic control—air-to-ground communications, navigation, as well as a computer-generated view out of the simulated cockpit. Ames used this facility to test, cheaply and quickly, all types of proposed improvements to cockpits and air traffic control systems.

Material scientists at Ames focused on aircraft fire safety. Data showed that the number of passengers who survived an aircraft fire was largely determined by their egress time and by the flammability of the aircraft seats. As a result, John Parker, an expert on foam-making, designed a seat of conventional urethane foam and covered it with a fire-blocking felt that was both fire resistant and thermally stable. In addition, the new seats were easy to manufacture and maintain, and were durable, comfortable and lightweight. In controlled fire tests done by the FAA on a C-133 and a B-720, passengers escaped the post-crash fires one minute faster than with earlier seats. Based on these tests, in October 1984 the FAA issued a new regulation on the flammability of aircraft seats. By October 1987, more than 600,000 seats were retrofitted, at a cost estimated at

A first flow calibration test in May 1986, for the 80 by 120 foot test section.

$22 million. Twenty-five lives each year that might be lost in fires were spared, it is estimated, because of these seats. John Parker and Demetrius Kourtides of Ames' chemical research project branch continued to work on fire-resistant materials, especially lightweight composite panels. Although this was an important application of materials science, it was mainly Ames' expertise in writing complex software that led to the expansion of its flight safety work for the FAA into the 1990s.

Upgrading the Wind Tunnels

Ames' wind tunnels still tied together its work in computational fluid dynamics at the start of aircraft design and automated flight testing at the end. Ames continued to invent new techniques to make more efficient use of its tunnels. With laser speckle velocimetry, for example, Ames solved the seemingly intractable problem of measuring unsteady fields in fluid flows. By seeding the air with microparticles, then illuminating it with a coherent light like that of a pulsed laser, they created speckled patterns which were superimposed on a photographic plate to create a specklegram. This specklegram recorded the entire two-dimensional velocity field with great spatial resolution. From this single measurement, aerodynamicists easily obtained the vorticity field generated by new aircraft designs. Similarly, in 1987, Ames' fluid mechanics laboratory started working closely with chemists at the University of Washington to develop pressure-sensitive paints that would turn luminescent depending on the amount of oxygen they absorbed. The paint was easily sprayed on an aircraft surface

Pressure sensitive paint on an F-18.

The NFAC at night.

before tunnel or flight testing and returned good data on the distribution of air pressure over the aircraft surface.

In addition to such upgrades in measurement equipment, Ames started upgrading what was still the world's best collection of wind tunnels. Ames had built many new special purpose tunnels in the 1950s and 1960s, but many of the general purpose tunnels built in the 1940s had started to degrade. In 1967 NASA participated in a nationwide review of American wind tunnels, and three at Ames were designated as being key national resources—the 40 by 80 foot, the 12 foot pressure and the Unitary. (The vertical motion simulator and the arc jet complex were designated national resources in their categories.) Ames then planned a long-term effort to bring these tunnels up to the state of the art, and to keep all of its tunnels operating safely. Of these efforts, perhaps the most significant was the December 1987 rededication of the National Full-Scale Aerodynamics Complex (NFAC).

The 40 by 80 foot wind tunnel, the largest in the western world since its opening in 1944, remained Ames' most unrivalled tunnel. It had been in almost constant use, and had saved engineers from making countless design mistakes. Ames people expected that any funds invested in updating it would be returned manyfold in better aircraft. For over a decade, Mark Kelly led groups from Ames to headquarters asking for funds for repowering the 40 by 80 foot tunnel and adding a new test section.

On 2 November 1978, Syvertson turned the first spade of dirt under the new 80 by 120 foot test section of the now renamed NFAC. (In addition to the one tunnel housing the two test sections, the complex also included Ames' outdoor

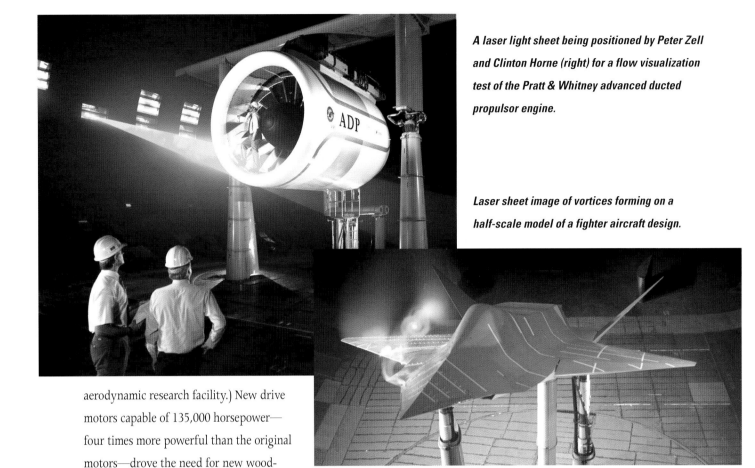

A laser light sheet being positioned by Peter Zell and Clinton Horne (right) for a flow visualization test of the Pratt & Whitney advanced ducted propulsor engine.

Laser sheet image of vortices forming on a half-scale model of a fighter aircraft design.

aerodynamic research facility.) New drive motors capable of 135,000 horsepower—four times more powerful than the original motors—drove the need for new wood-composite fan blades and minor strengthening of the hull. The 40 by 80 foot section would continue to work as a closed-loop tunnel, with an air circuit a half mile long. The 80 by 120 foot tunnel would be open at both ends, rather than closed loop, which reduced the cost to $85 million and construction time to an additional six months. It would gulp in air through a horn-shaped inlet as big as a football field. Kenneth Mort, lead aerodynamicist on the upgrade, built a 1/50th scale model of the tunnel itself to show that Bay Area winds would not unacceptably degrade the smooth flow of the test air. This bigger section would operate at an airspeed only one-third that of the airspeed in the 40 by 80 foot test section, but was big enough to evaluate ever-larger military and commercial aircraft. Furthermore, the higher speed and larger size of the modified facility made it ideal for Ames' growing body of work in VTOL aircraft, helicopters and aeroacoustics. The larger test section minimized tunnel-wall interference, which

Scale model of the NFAC, used to study the complex airflows through the tunnel before construction began to add the 80 by 120 foot test section.

Drive fans for the 80 by 120 foot wind tunnel during reconstruction.

Insets: Damage to the fan blades in the 1982 accident.

worsened at low speeds or when air was deflected downward and outward by rotorcraft. Since sound waves took some distance to propagate, large test sections were also important in aircraft noise studies, an issue becoming more politically sensitive. To make the new tunnel better suited to aeroacoustic research, and to reduce the noise made while the tunnel was running, Ames engineers lined the test sections with six inches of sound-absorbing insulation. Cranes were added for moving around larger models. Better sensors, model mounts, wiring and computers were added for data collection. Construction of the composite tunnel ended in June 1982.

Just before noon on 9 December 1982—with only two months of shakedown tests to go before it would be fully operational—the NFAC suffered a serious accident. While running at 93 knots in the 80 by 120 foot test section, close to its maximum speed, a slip joint holding the hinge mechanism on vane set number 5 slipped. The entire lattice work of vanes broke up and its debris was blown into the drive fans. Vane set 5 stood 90 feet high, 130 feet wide, and weighed 77 tons. Located 100 feet upwind of the fans, the nose sections of the vanes hinged to guide airflow around a 45 degree corner from the new 80 by 120 foot section into the old

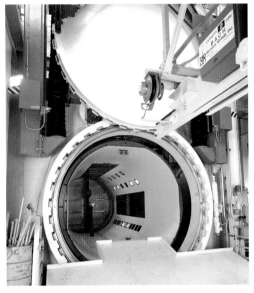

The air lock and test section of the 12 foot pressure tunnel.

tunnel. All ninety fan blades, carefully handcrafted of laminated wood, were destroyed. The institutional trauma of the accident announced itself with a terrifying thump heard around the Center. The accident affected morale throughout Ames though Syvertson assumed the blame that, as Center director, was ultimately his to bear. Ames had done a poor job supervising design and construction of the vane set. More stunning, Ames could no longer be proud of its safety record (though no one had been hurt in this accident). Syvertson had earlier nominated the Ames machine shop for a NASA group achievement award to recognize its year of no loss-time accidents. When NASA headquarters refused the nomination, on the grounds that NASA gave no awards for safety, Syvertson was so incensed that he refused the NASA Distinguished Service Medal that he was to be awarded.

The 20-blade axial flow fan that provides airflow through the 12 foot pressure wind tunnel.

Yet Ames wrested success from the tragedy. Ames tunnel managers shuffled the test schedule to make use of smaller tunnels, so that the accident added little to the two-year backlog of tests waiting for the tunnel to open. Ames estimated it would take one year and cost $13 million to repair. However, a blue ribbon panel of aerospace experts convened by NASA and led by Robert Swain suggested taking this opportunity to make additional upgrades to boost the NFAC's reliability. This raised the total renovation cost to $122.5 million, the amount Ames had originally requested. Better instrumentation, stronger structural steel, and turning vanes with

Boeing 737 being tested in air chilled by Freon, in the 12 foot pressure tunnel.

sophisticated airfoils and no movable parts all created a more capable tunnel. New wiring for 1,250 channels pushed data at rates up to two million bits per second into computers where they could be instantly compared with theoretical predictions. Although both tunnels could not be run at the same time, engineers could set up tests in one tunnel while the other one ran. On 26 September 1986, the Ames project group led by Lee Stollar started the first preliminary tests. Almost a year passed before the NFAC was declared fully operational.

Following the upgrade, airspeeds in the 40 by 80 foot test section could reach 345 miles per hour, the low cruise speed for many aircraft. The 80 by 120 foot tunnel, operating at 115 miles per hour, became the world's largest open-circuit tunnel. It proved

The rebuilt 12 foot pressure tunnel, June 1995.

Atmosphere of Freedom   Sixty Years at the NASA Ames Research Center

The vertical motion simulator at rest, being prepared for another mission.

The renovated vertical motion simulator, in a time-lapse photograph.

especially useful in studies of actual aircraft and in situations where low speed handling was especially critical, like during landing and takeoff. It has been used to test a variety of aircraft on a large scale—fighter jets, lifting-body configurations, Space Shuttle models, supersonic transports, parachutes, and even trucks and highway signs.

Once Ames got the tunnel renovation program back on track after the accident, it focused on the 12 foot pressure tunnel. The tunnel hull had, since its opening in 1946, undergone constant expansion and contraction as it was pressurized to achieve its extraordinarily smooth flows of air and then depressurized. In December 1986, such extensive, unrepairable cracks in the welds were discovered during a detailed inspection that Ames decided to rebuild the hull completely. Models of virtually every American commercial airliner had been tested in the 12 foot pressure tunnel, and aircraft designers hoped to continue to rely upon it. Beginning in 1990, a project team led by Nancy Bingham stripped and rebuilt the closed-loop pressure vessel, and installed an innovative air lock around the test section. The new air lock let engineers enter the test section without depressurizing the entire tunnel, boosting its productivity and reducing the pressure cycling that had previously degraded the hull. Ames also integrated new test and measurement equipment, and upgraded the fan drive. The 12 foot pressure tunnel was rededicated in August 1995, creating a superb test facility at a renovation cost of only $115 million.

Visualization of a transonic flow field, showing the shock wave shape. A reconstructed laser hologram was used to make this interferogram image.

The 20 g centrifuge, opened in 1962, was upgraded in 1993.

The 20 g centrifuge was built underneath the 40 by 80 foot tunnel in 1965 to test how well experiments flown in Biosatellite would survive the hypergravity of takeoff and landing. By the early 1990s, it was one of six hypergravity facilities at Ames, but the only human-rated centrifuge in NASA. "It's a simple facility," noted centrifuge director Jerry Mulenburg, "but it's very flexible for our purposes."[3] Ames upgraded its controls and data collection system, completed in March 1994, and built a new treadmill cab to fit on the end of its 58 foot diameter arm for exercise tests in it up to 12.5 g forces.

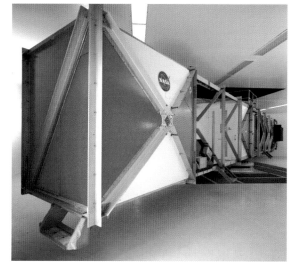

A major upgrade of the vertical motion simulator (VMS) was completed in May 1997, with construction of a new interchangeable cockpit. Ames built the new T cab in-house, specifically to satisfy the needs of NASA's tilt rotor and high speed airliner programs. The new T cab had a side-by-side arrangement and an all-glass cockpit, so pilots could press easily altered touch-screens rather than actual instruments. The 270 degree view out the window was twice that of the other four cabs available to the simulator, which simulated helicopters, airplanes and the Space Shuttle.

Since being placed in service in 1955, the Unitary plan wind tunnel, like most Ames facilities, had been in almost constant operation. Such constant operation was planned, since Ames had designed the tunnel with massive diversion valves that allowed a test to be run in one section while models were set up in the other two. The drive system had accumulated over 70,000 hours of operation, as the Unitary complex tested every military aircraft, every significant commercial transport, and every manned spacecraft

Diverse Challenges Explored with Unified Spirit: 1969 – 1989

Diffuser and contraction vanes during modifications to the Unitary plan wind tunnels.

since its inception. The 11 foot transonic tunnel still had a 2.5 year backlog of tests, and the cost had risen to $300,000 for a one-week test. Ames shut down the Unitary in 1996 for an $85 million renovation to make it operate more efficiently. Modernization would automate the control system and improve flow quality in the transonic section by adding honeycomb flow straighteners, turbulence reduction screens, and segmented flaps in the wide-angle diffuser to eliminate flow separation. The Unitary modernization completed the overhaul of Ames' most valuable physical assets and provided the research tools needed to continue moving aircraft concepts to flight tests—as it had with VTOL aircraft.

VERTICAL TAKEOFF AND LANDING AIRCRAFT

The separation of lift from thrust (that is, using an airfoil and an engine instead of flapping wings) was the insight that made powered flight possible. Reuniting lift and thrust into propulsive lift, with the new technology earned over a half century of flight, promised a revolution in the relationship between aircraft and the populations they serve. Wing-tip rotors lift the aircraft like a helicopter, then the rotors tilt forward like propellers and transfer the lift from the rotors to the airfoil until the aircraft flies like an airplane. Helicopters do not fly forward efficiently. Fixed-wing aircraft find forward efficiency in higher wing loading, which requires longer runways, which then mandate bigger and more congested airports, farther from population centers. Tilt rotors can fly longer distances than helicopters, yet require little more space than a helipad to takeoff and land.

Following World War II the Transcendental Company, a small American firm, built their Model 1-G tilt rotor which flew

Ames aerodynamicists tested a wide variety of VTOL aircraft and helicopters during the 1960s to establish a base of research data. Here the Hiller rotorcycle YROE-1, made by Hiller Helicopter in nearby Palo Alto, California, hovers in front of the Ames hangar.

Flight testing of the Bell XV-3 Convertiplane.

100 flights for a total of 23 hours. The Model 1-G was very small (1,750 pounds), was never fully converted to forward flight, and crashed in 1954 to end the project. It proved the concept, but it did not end debate over which vertical takeoff and landing configuration minimized the weight penalty—that is, minimized the need for more powerful engines and stronger shafting that made propulsive-lift aircraft heavier than regular airplanes. So the U.S. Army let three contracts—to McDonnell Aircraft for the XV-1, to Sikorsky for the XV-2 stoppable rotor, and to Bell Aircraft Corporation for the XV-3.

Bell had started working on tilt rotors in 1944, and accelerated their research by hiring Robert Lichten, an engineer for Transcendental. For the next two decades, Lichten would be the dominant player in American tilt rotor development. The XV-3 that Lichten and Bell designed for the U.S. Army was a small aircraft, only 5,000 pounds gross weight. A single engine mounted in the center turned a complex gear box that powered large rotors at the tips of the wings.

The XV-3 first flew in 1955, and every flight was nerve racking. The cockpit vibrated up and down whenever it hovered. To compensate for an engine simply too underpowered, Bell built the airframe too light. In a hover flight, in 1956, a rotor pylon coupling failed catastrophically and the pilot was severely injured. Bell strengthened the structure, thus restricting it to ground-tethered flights while they searched for solutions.

Diverse Challenges Explored with Unified Spirit: 1969 – 1989

Following this crash, Ames engineers entered the picture in 1957, and started with some tests in the 40 by 80 foot wind tunnel. The XV-3 flew again in 1958, with NASA pilot Fred Drinkwater at the controls to define the conversion envelope between vertical and horizontal flight. Full conversion from helicopter mode to conventional forward flight was flown in August 1959, and the entire XV-3 test program proved a major advance in understanding the transition from ground to air. The XV-3 program ended in 1965 after a rotor pylon tore loose from the XV-3 while it was inside the 40 by 80 foot tunnel. For a few months, Ames and Bell engineers did a radical redesign of the remaining pylon to test ways to improve pylon stability—a major weak link in tilt rotor design. In 1966, Ames finally mothballed the XV-3.

X-14B VTOL aircraft over the San Francisco Bay.

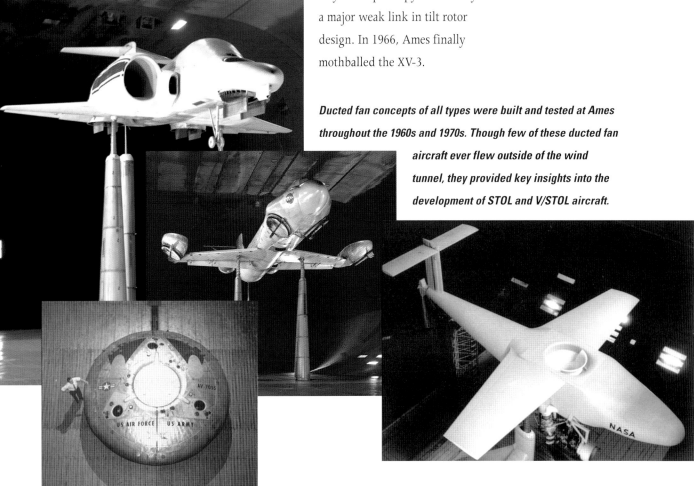

Ducted fan concepts of all types were built and tested at Ames throughout the 1960s and 1970s. Though few of these ducted fan aircraft ever flew outside of the wind tunnel, they provided key insights into the development of STOL and V/STOL aircraft.

AV-8B Harrier during a precision hover test at the Crows Landing Auxiliary Airfield near Moffett Field.

AV-8B Harrier and V/STOL Handling Characteristics

In the 1960s, though, the excitement over propulsive lift swirled around vectored-thrust jet aircraft. NASA contracted with British Aerospace to build the XV-6A Kestrel, which flew so well that it was quickly redesigned into the Harrier, known in the United States as the AV-8B. The jet exhaust nozzle of the Harrier was pointed downward to lift it off the ground, then rotated backward to provide forward thrust. The Harrier's efficiency was poor when hovering, but it otherwise performed well in the marine fighter/attack role. Ames was fortunate to receive early prototypes of the Harrier, which they put in the 40 by 80 foot wind tunnel to gain a better understanding of the very complex airflows of vectored thrust.

X-5B aircraft hovering at Ames, August 1969.

Ames also used their flight tests with the AV-8B Harrier, as well as wind tunnel and simulator tests, to author handling qualities definitions for all future V/STOL aircraft (for vertical and short takeoff and landing). V/STOL aircraft feel different to any pilot, whether they train on helicopters or fixed-wing aircraft. First published as a NASA technical note, these handling quality definitions were applied to all V/STOL aircraft in NATO and in the U.S. military through its V/STOL flying qualities specification.

But ideas for higher-efficiency propeller-driven V/STOL aircraft continued to percolate. NASA let contracts for a variety of approaches—like the Ryan XV-5A which used turbine driven lift fans. For the U.S. Army, Vought (later LTV) built several XC-142 tiltwing prototypes, which flew well but were very complex and had problems in conversion.

Diverse Challenges Explored with Unified Spirit: 1969 – 1989

The XV-15 in hover flight.

Bell invested its own money, with considerable help from Ames, in designing its Bell Model 300. It had good hover and rotor efficiency and its pylons proved stable in 40 by 80 foot tunnel tests. Ames had worked hard, since the demise of the XV-3, to solve the lingering problems of tilt rotor aerodynamics.

In 1970, NASA decided to fund another effort in tilt rotors. Foreign competitors were especially strong in small aircraft and helicopters, and NASA headquarters wanted America to regain the lead through a technological leap. In the debate that ensued, aerodynamicists at Langley favored a tilt-wing approach. But C. W. "Bill" Harper, then director of aeronautics at NASA headquarters, sided with his former colleagues at Ames in favoring the tilt-rotor approach.

XV-15 Tilt Rotor

A key factor in Ames getting the XV-15 project was its close relationship with the Army Aviation Research and Development Laboratory, co-located at Ames since 1965. Because of this alliance with the Army, Ames had funds to refurbish an inactive 7 by 10 foot tunnel for small scale tests in advance of tests in the 40 by 80 foot tunnel. The complex aerodynamics of helicopters and VTOL aircraft meant that they had to be tested in full-scale tunnels. On VTOLs, effects could not be scaled, interference from downwash was extreme, and the hard work was in the details. The XV-15 was designed for medical evacuation and search and rescue missions that the U.S. Army had encountered during the war in Vietnam. The XV-15 had a gross weight of 15,000 pounds, a payload of 4,000 pounds, a cruising speed of 350 knots, and a range of 1,000 nautical miles—roughly twice that of the best helicopters. In 1970, management of the XV-15 went to a joint

XV-15 with advanced tilt rotor blades, in a hover acoustic test observed by Paul Espinosa and Doug Sanders, December 1990.

XV-15 tilt rotor in conversion to forward flight.

NASA-Army project office with David Few in charge. Half of the $50 million required for the project came from Ames, half from the Army. Hans Mark gave it his full support. This was the first time Ames bought an aircraft meant to be a full-scale technology demonstrator—to show the military and airlines how easily they could build such an aircraft for regular service.

In September 1972, the Ames/Army project office gave both Bell and Boeing $500,000 design contracts, and in April 1973 they declared Bell the winner. Led by program manager Ken Wernicke, Bell then apportioned the work for two XV-15 prototypes using standard components as much as possible. Rockwell fabricated the tail assemblies and fuselage, Avco-Lycoming modified a T-53 engine, and Sperry Rand designed and built the avionics. Ames aerodynamicists immediately started modelling wind flows around the aircraft, for example, formulating equations to predict whirl flutter caused by a rigid rotor spinning on a pylon.

In exterior configuration, the XV-15 differed little from the XV-3. But as happens so often in aircraft development, better propulsion made the whole system remarkably better. The Lycoming turbine engines had much better power-to-weight ratios than those on the XV-3. Bell mounted one at each wing tip to turn the three-blade proprotors, which were 25 feet in diameter. The only

JVX rotor blade mounted for testing at the outdoor aerodynamic research facility (OARF). Ames opened the OARF in 1979 specifically to check out models before they are installed in the larger wind tunnels, to study balance and gas reingestion on tilt rotors, and to obtain acoustic data on all varieties of aircraft.

Diverse Challenges Explored with Unified Spirit: 1969 – 1989

XV-15 tilt rotor, in takeoff mode, in the 40 by 80 foot wind tunnel.

cross-shafting in the XV-15 was designed to carry loads only when one engine failed.

The first XV-15 prototype rolled out of the hangar on 22 October 1976 for ground tests by Bell pilots. On 3 May 1977, Bell chief project pilot Ron Erhart first flew the XV-15: "It flew just like the simulator," wrote Erhart, "but with better visuals."[4] On 23 March 1978, the XV-15 arrived at Ames for a more intensive series of flights. Ames pilots tested it in engine-out flight, and found the cross-shafting worked well in an emergency. On 24 July 1979, it made the full conversion from vertical to forward flight.

Ames discovered some fascinating aerodynamic problems. When the proprotors were tilted at certain angles relative to the wings, a large vortex was generated over each wing that caused strong buffeting in the tail. The only solution was to brace and stiffen the tail. Pilots found it took some time to get the feel of the conversion, and that it behaved oddly during taxiing and in light wind gusts.

In spring 1980 Ames opened its outdoor aerodynamic research facility (OARF), essentially a tilt rotor tie-down facility on a hydraulic lift. By raising the wheel height from two to fifty feet off the ground (to accommodate the large proprotors) they could evaluate the XV-15 flying through air in any flight configuration. Ames aerodynamicists could measure rotor torque, fuel consumption, aircraft attitude, pilot control and—at various hover altitudes—ground effects, downwash, handling qualities, exhaust gas reingestion, zero wind force and moment data, and noise levels.

The XV-15 program was scientifically interdisciplinary—human factors, computing and digital controls all helped out in the crucial area of pilot workload. Flight data were cross-checked with tunnel data, which were matched to the formative efforts of computational fluid dynamics. The XV-15 culminated in an intense research program at Ames to further develop the VTOL concept and to prove its commercial value and military utility. Yet it took some extraordinary steps to move the tilt rotor to its next iterations.

A terrain model manufactured by Redifon, Ltd., opened at Ames in 1971. Coupled via video to a cockpit simulator, it generated visual cues for research on short takeoff and landing aircraft.

V-22 Osprey

In 1978 Ames, emboldened by Hans Mark's duty as secretary of the Air Force, directly, and without success, tried to get the Army or Air Force to buy an improved tilt rotor for search and rescue missions. Mark made a special, and again unsuccessful, pitch to Admiral Holloway, former Chief of Naval Operations who led the investigation into the failed April 1980 effort to extract the American hostages from Iran. Resistance came because the U.S. Air Force had always fought its air wars from protected airfields, and thus saw no need for an operationally independent aircraft. And the Army already had expensive new helicopters entering service to fly those same missions.

Mark moved from the Pentagon to be deputy administrator of NASA early in 1981, and one of his first decisions was to support Ames' efforts to take the XV-15 to the Paris Air Show. It was a hit. The new secretary of the Navy, John Lehmann, saw it at the show and became a staunch advocate of the tilt rotor. In 1982, NASA departed from usual practice and let its experimental aircraft be used in operational tests. The Army flew the XV-15 to simulate electromagnetic warfare near Fort Huachuca, Arizona. The Navy evaluated it aboard the USS *Tripoli*. P. X. Kelley, commandant of the Marine Corps, also became a tilt rotor advocate, especially after the 1982 Argentine-British conflict over the Falkland Islands. Missiles used in the conflict showed that standoff distances between ships and a hostile shore had to be farther than the short operating ranges that ship-based helicopters allowed.

In 1983, the Marines issued the specification for what became the V-22 Osprey, a VTOL designed to replace the Boeing Vertol CH-46 and the Sikorsky CG-53 assault helicopters. Bell Helicopter Textron Inc. of Fort Worth teamed with Boeing Vertol of Philadelphia and won the contract in 1985. The V-22 was three times the size of the XV-15, with a total gross weight of 40,000 pounds, but otherwise similar. It would carry 24 heavily armed Marines from ship to shore in amphibious assaults. In a significant advance in airframe technology, many of the key structural members of the V-22 were made of fiber-reinforced graphite-epoxy laminate. The V-22 designers were comfortable using composites so extensively because of the VTOL technology database developed at Ames, and overseen by John Zuk, Ames' chief of civil technology

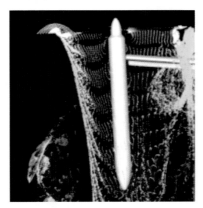

Computer simulated images of viscous flow about rotor and wing of the V-22 Osprey in hover.

programs. The first V-22 flew on 19 March 1989, though it continued to work itself slowly into military service.

The success of the V-22 in military service should pave its way into civil transport, where tilt rotors are most needed. Commuter airlines now flying

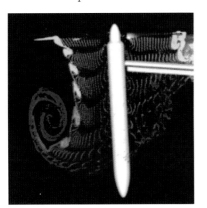

small, propeller-driven Brazilian Embraers or European Fokkers may find that forty seat tilt rotors, operating independent of congested airports, could move people much faster door-to-door. Ames led a study funded by the FAA, NASA, and DoD on the potential of the Osprey for civil transport, and the New York Port Authority asked Ames to help explore the potential of tilt rotors to solve local transportation problems.

JET-STOL AIRCRAFT

Ask pilots, and they'll say that just as important as flying fast, is being able to fly slowly. Slow-speed flight remained out of fashion as engineers built aircraft to go faster and farther, but Ames researchers always held a great deal of respect for complex airflows at slow speeds. So Ames developed

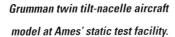

Grumman twin tilt-nacelle aircraft model at Ames' static test facility.

expertise in aerodynamics at slow speed in order to help in the design of aircraft that handled better in the trickiest parts of any flight—takeoff and landing. Better performance at slow speeds also resulted in aircraft that could take off or land on much shorter runways—important for commuter airlines operating from smaller regional airports or for military pilots operating from unimproved foreign airfields.

Thus, in conjunction with researchers from the U.S. Army, Ames used its expertise to build a series of STOL aircraft (for short takeoff and landing) like the augmentor-wing quiet short-haul research aircraft (QSRA), the rotor systems research aircraft (RSRA), and the E-7 short takeoff and vertical landing (STOVL) test model.

Augmentor Wing STOL, Quiet Short-Haul Research Aircraft

Ames first worked to develop specific technologies that airframe companies could apply to other short takeoff and landing aircraft. A rotating cylinder flap, for example, improved lift by energizing boundary layers as it turned airflow downward over the trailing edge of the wing. Ames installed a rotating cylinder flap on an OV-10 Bronco and, even though radically modified, the OV-10 proved the point faster and cheaper than building a completely new technology demonstrator. Ames shortened the wings, removed the flaps and pneumatic boundary layer control, shortened the propellers, boosted the gross weight from 8,500 to 11,500 pounds to get rotation to the cylinders, and crossshafted the two engines for better performance at slow speeds. Before its first flight in August 1971, Ames completely tested the OV-10 in the 40 by 80 foot tunnel. The rotating cylinder used so little power that full horsepower was available for takeoff. Compared with the basic OV-10, it achieved 33 percent better lift.

In the 1970s, Ames and Canadian researchers joined to

V-22 Osprey in transition.

Diverse Challenges Explored with Unified Spirit: 1969 – 1989

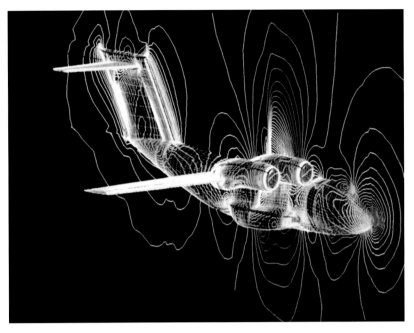

Computer image of the Japanese Asuka STOL.

study jet STOL with a complete flying test bed. They modified a government surplus deHavilland C-8 Buffalo turboprop aircraft to demonstrate the technology of powered-lift ejector augmentation. The modified Buffalo first flew on 1 May 1971 and remained at Ames in flight tests through 1976. Its thrust-augmentor wing achieved augmentor ratios of 1.2 with significant gains in lifting coefficients, so that it could fly as slow as fifty knots and approach the landing field at sixty knots. It routinely demonstrated takeoffs and landings in less than 1,000 feet, with ground rolls less than 350 feet. After a full range of technical flight tests, Ames pilots flew the Buffalo in a series of joint flights—with the FAA and the Canadian department of transportation—to develop certification criteria for all future powered-lift aircraft.

Ames' next iteration of powered-lift aircraft was the QSRA (for quiet short haul research aircraft). Boeing of Seattle built the QSRA from the C-8 Buffalo and four spare Lycoming turbofan engines. The engines were mounted on top of the wing, so that the exhaust air blew over the upper surface, creating more lift, while the wing shielded the noise from the ground below. The QSRA wing was also entirely new, emulating a supercritical airfoil capable of Mach 0.74 (though the QSRA never went that fast) and a

The quiet short-haul aircraft (QSRA), a highly modified C-8A, undergoing carrier trials on board the USS Kitty Hawk near San Diego.

wing loading of eighty pounds per square foot. The result was a very quiet, efficient aircraft, capable of very short takeoffs and landings.

Boeing delivered the QSRA to Ames in August 1978, and it quickly validated the concept of upper-surface blowing. The QSRA could fly an approach at only sixty knots, at a steep, twenty degree angle. "It feels as if it's coming down like an elevator," said Jim Martin, QSRA chief test pilot.[5] During carrier trials in July 1980 aboard the USS *Kitty Hawk*, with wind over the deck at thirty knots, the QSRA took off in less than 300 feet and landed in less than 200. In zero wind conditions, during Air Force tests to simulate operations on bombed runways, the QSRA took off in less than 700 feet and landed in less than 800 without thrust reversers. The real military payoff, however, was that augmented lift boosted payload capacity by 25 percent. In 1983, Jim Martin and Robert Innis flew the QSRA to the Paris Air Show to encourage companies to use the technology in commuter aircraft. Short takeoffs and landings were important to operating bigger aircraft on smaller, local runways; more important, the QSRA far surpassed federal requirements for noise abatement. It flew a demonstration landing into the Monterey, California, airport completely undetected by the airport monitoring microphones.

North American OV-10A Bronco rotating cylinder flap aircraft powered by two interconnected T-35 turbine engines.

Over the fifteen years that Ames pilots flew the QSRA, they conducted 697 hours of flight tests which included more than 4,000 landings—averaging nearly six landings per flight hour. More than 200 research reports emerged from data collected on the QSRA. Once the aircraft itself was understood, the Ames QSRA team, led by John Cochran and then Dennis Riddle, used it more as a test bed for new technologies. The renamed NASA Powered-Lift Flight Research Facility provided an ideal platform, beginning in November 1990, to test a jump-strut nose gear that kicked up an aircraft nose during takeoff. Ames retired the QSRA in March 1994.

A fuel-efficient turboprop model undergoing early tests in the 14 foot wind tunnel, 1980.

Rotor Systems Research Aircraft

Another unusual aircraft that bridged the worlds of vertical and fixed-wing flight was the rotor systems research aircraft (RSRA). Sikorsky built two RSRAs, originally for research at Langley, that arrived at Ames in September 1979. A NASA/Army team designed them as flying wind tunnels—highly instrumented, flying test beds for new rotor concepts. One was built in a helicopter configuration, powered by two turboshaft engines. The second had a compound configuration, meaning that it could fly with lift provided by two short wings as well as by the helicopter rotor. Two turbofans were added as auxiliary engines, and the aircraft was instrumented to measure main and tail rotor thrusts and wing lift. Warren Hall served as RSRA project pilot.

The helicopter RSRA was later modified to test an X-wing configuration proposed by the Defense Advanced Research Projects Agency (DARPA). The X-wing RSRA had a single rotor with four blades, built out of composite materials, that lifted the aircraft vertically like a helicopter. Air blown through a fore or aft strip along each rotor blade provided pitch and roll control. As its turbojet engines thrust it forward as fast as Mach 0.8, the rotor provided lift as a

Rotor System Research Aircraft.

Sikorsky bearingless rotor, undergoing tests in the 40 by 80 foot wind tunnel.

symmetrical airfoil with an X shape. The convertible engine divided its power as it shifted between rotor flight and jet exhaust. In aircraft mode, the air blown through the rotor blades provided lift and control. The RSRA flew only three times in the X-wing configuration, before being abandoned as too difficult to control.

Rotary Wing Aircraft

Ames began working on rotorcraft in the early 1970s as its research relationship with the Army aeroflightdynamics directorate expanded. Initially, studies focused on pilot control during terminal operations—getting aircraft on and off the ground, especially during bad weather—and Ames built a sophisticated series of flight simulators for helicopter pilots.

Ames' inventory of rotorcraft jumped in the late 1970s, when four other helicopters were transferred to Ames from Langley: the UH-1H and AH-1G for rotor experiments, and the SH-3 and CH-47 for operational studies. Ames established a new helicopter technology division to focus on these aircraft, to pursue research in rotor aerodynamics and rotor noise, and to develop new helicopter technologies. The Army, likewise, continued to beef up the technical expertise in its aeromechanics laboratory, led by Irving Statler. Ames and Army aerodynamicists developed a free-tip rotor, for example, with a tip that was free to pitch about its own axis, which was forward of the aerodynamic center.

Ames and the Army Aeromechanics Laboratory opened this 21 by 31 centimeter water tunnel in 1973, to provide better visualization flows around oscillating airfoils.

Diverse Challenges Explored with Unified Spirit: 1969 – 1989

Ames' anechoic chamber, in October 1995, set up for a DC-10 acoustic array calibration. Ames and the U.S. Army Aeromechanics Laboratory opened this chamber in 1978 to unravel the complexities of rotor noise. It was designed so that neither noise nor air bounced around inside the chamber during a test. From the earliest experiments on UH-1H model rotor blades, Ames explored the discrepancies between test results and linear acoustic theory.

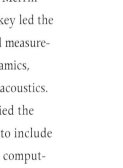

Ames built a model that showed that the free tip rotor reduced power at cruise speed, minimized vibratory flight loads, and boosted lift by sixteen percent.

Ames flew the UH-1H to develop automatic controls for landing a helicopter, culminating in a fully automatic digital flight guidance system known as V/STOLAND. Principal engineers George Xenakis and John D. Foster first developed a database of navigation and control concepts for instrumented flight operations. Kalman filtering extracted helicopter position and speeds from ground-based and onboard sensors. To define the helicopter's approach profile and segregate it from other airport operations, the system investigated several helical descending flightpaths. Lloyd Corliss then led a series of UH-1H test flights on flying qualities for nap-of-the-earth operations, and Victor Lebacqz used it to devise certification criteria for civil helicopter operations. Later, project pilots Dan Dugan and Ron Gerdes flew the UH-1H in the first demonstration of automatic control laws based on the nonlinear inverse method of George Meyer.

The Bell AH-1G White Cobra arrived and was highly instrumented for the tip aerodynamics and acoustics test (TAAT) to establish better prediction methods for this type of twin-blade rotor. Ames got the highly instrumented rotor blades that the Army had used for its operational loads survey and added additional absolute pressure instrumentation to the rotor tips. Thus, one rotor blade returned 188 pressure transducer measurements, with 126 more measurements added by the other blade and the rotor hub. Robert Merrill was chief pilot and Gerald Shockey led the project, which returned detailed measurements of aerodynamics, performance and acoustics.

Ames modified the CH-47B Chinook to include two digital flight comput-

UH-60 Blackhawk outfitted for a blade-vortex noise experiment.

CH-47B Chinook helicopter hovering over the Ames ramp.

ers, a programmable force-feel system, and a color cathode-ray tube display. This system allowed wide variations in the helicopter's response to pilot controls, making it an ideal variable stability research helicopter. Ames used it in flight simulations to define new military handling qualities. In close cooperation with Stanford University researchers, Michelle Eshow and Jeffery Schroeder used the CH-47B to investigate multiple input and output control laws developed on Ames' vertical motion simulator. The Army let Ames use the CH-47B from 1986 until September 1989, just before they closed out the line that remanufactured them into a CH-47D suitable for Army duty.

To carry forward this variable stability research, in 1989 Ames acquired a Sikorsky JUH-60A Blackhawk. Known as RASCAL (for rotorcraft aircrew systems concepts airborne laboratory), it carried extensive vehicle and rotor instrumentation, a powerful 32-bit flight control computer, and image generators for the cockpit. "We're putting a research laboratory in a helicopter," said RASCAL program manager Edwin Aiken. "Now when we experiment with flight control software, advanced displays or navigation aids, we can get a realistic sense of how they work."[6] Ames and Army engineers used RASCAL to develop a range of new technologies—active sensors like millimeter wave radar, passive sensors

UH-1H helicopter equipped with V/STOLAND, a digital avionics system that evaluated flight performance through various configurations of automatic control, display, guidance and navigation.

Diverse Challenges Explored with Unified Spirit: 1969 – 1989

A visualization test of the flows off the tail rotor of an AH-1G Cobra helicopter model in the 7 by 10 foot wind tunnel operated by the U.S. Army, November 1973.

using infrared, and symbologies for advanced displays. The goal was to make helicopters respond to pilot controls with more precision and agility, to provide better obstacle avoidance and automated maneuvering close to the terrain, and to improve vehicle stability when carrying loads or using weapons. For example, Ernest Moralez helped devise algorithms that would automatically protect a flight envelope in which pilots could then maneuver freely.

Another UH-60 Blackhawk also entered the Ames inventory in September 1988 as part of the modern rotor aerodynamic limits survey (MRALS). Sikorsky Aircraft built two highly instrumented blades for the Ames/Army program. A pressure blade with 242 absolute pressure transducers measured air loads—the upward force produced as the blades turn. A blade with a suite of strain gauges and accelerometers measured the structural responses to air loads. The pressure blade alone returned a 7.5 megabit data stream, which demanded a bandwidth well beyond the state of the art. An Ames group, led by Robert Kufeld and William Bousman, devised a transfer system that returned thirty gigabytes of data during test flights in 1993 and 1994—data then archived onto optical laser disks in a jukebox storage system for immediate access via modem by rotorcraft designers. The UH-60 studies ended a ten-year air loads program, launched in 1984 and completed for only $6 million. Its legacy was an air loads database actively used to refine helicopter design and to better predict performance, efficiency, airflows, vibration and noise.

By taking novel technical approaches to first isolating and then solving seemingly intractable problems, and integrating their use of computa-

RASCAL helmet mounted display.

UH-60 RASCAL with infrared camera mount.

tion, tunnel and flight testing, Ames bolstered the core technologies found in all helicopters. Ames people made similar contributions to the Space Shuttle program. While other NASA Centers led systems design, integration and management, Ames tackled the tough issues of aerodynamic configuration and thermal protection.

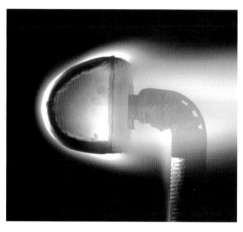

Test on ablative ceramic in the Ames 60 megawatt arc jet, in development of new materials for entering planetary atmospheres.

SPACE SHUTTLE TECHNOLOGY

In 1971, Ames established a small Space Shuttle development office, led by Victor Stevens, to coordinate all the people at the Center who were working on Shuttle technologies. Using the NFAC, the Unitary and 3.5 foot hypervelocity tunnels, Ames did half of all tunnel tests—to increasing speeds— during the crucial phase B of the Shuttle design. Ames people used the expertise earned in lifting body studies to refine the Shuttle configuration, and expertise earned in digital fly-by-wire to design controls for the Shuttle. Shuttle trainees spent fifty weeks in the Ames vertical motion simulator studying handling qualities during landing. Furthermore, Ames managed NASA's Dryden facility which served as the primary test facility and landing site for all early Shuttle flights. Despite the magnitude of these efforts, Ames worked on Shuttle technologies, as it had on Apollo technologies, without having the program dominate the mission of the Center. And as with Apollo, Ames' primary contribution was solving the problems of reentry and materials that got the Shuttle astronauts home.

Advanced space shuttle thermal protection materials in a plasma stream during arc jet tests.

When the Space Shuttle orbiter Columbia first touched down at Ames–Dryden in April 1981, shuttle commander John Young exited the

Diverse Challenges Explored with Unified Spirit: 1969 – 1989

Oil painting of the Space Shuttle tile team at work, in 1980, in the orbiter processing facility at Kennedy Space Center.

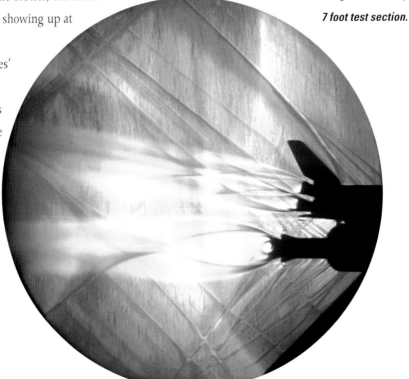

Space Shuttle plume imaged in the 9 by 7 foot test section.

orbiter, walked underneath, looked around, gave a thumbs up, then jumped with joy. The thermal protection system was the key to making the Space Shuttle the world's first reusable reentry vehicle. Heat shields used earlier on Apollo and other capsules had been rigid, with ablative materials designed to burn up while entering the atmosphere only once. The airframe of the Shuttle orbiter, however, would be flexible like an aircraft, with complex curves, and had to be built from a system of materials that rejected heat without ablating. Once NASA had decided, in the mid-1960s, on reusable insulation for the Shuttle orbiter, the airframe firms that hoped to build it started showing up at Ames for advice and tests.

Howard Larson took over Ames' thermal protection branch in 1968. Larson had spent most of the 1960s studying how ablation changed the shape of bodies that entered Earth's atmosphere—like meteors, ballistic missiles and capsules— and thus affected their aerodynamic stability. Nonablative thermal protection, however,

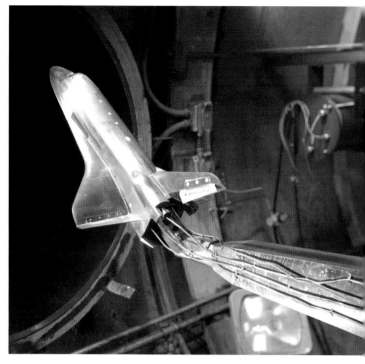

Rockwell International tested this design of the Space Shuttle orbiter in May 1973 in the 3 by 5 foot hypersonic wind tunnel.

required an entirely new class of heat shield materials. To help evaluate these, in 1970 Larson hired Howard Goldstein, a thermodynamicist and materials scientist then running arc jet tests at Ames for a NASA contractor. As the pace of materials testing accelerated, the Shuttle contractors increasingly bumped up against the size and run-time limitations of Ames' 20 megawatt arc jet. But Ames still had the largest direct-current power source in NASA, as well as an enormous infrastructure for compressing atmospheres. In 1971 Dean Chapman, who as director of astronautics oversaw Larson's work, secured funds to build a 60 megawatt arc jet. Materials science quickly took on new prominence at Ames.

In 1971 Ames directed its efforts to help Johnson Space Center evaluate a new class of reusable surface insulation for the Shuttle. Lockheed Missiles and Space had developed tiles based on low-density rigid silica fiber—called the LI-900 tile system—that was selected in 1973 to cover two-thirds of the Shuttle's surface. Goldstein led Ames' effort to apply the database built during arc jet tests of this and other candidate materials to develop improved heat shields. An early Ames product was a black borosilicate coating (called RCG for reaction-cured glass), that provided a lightweight and easily manufactured surface

Space Shuttle orbiter model undergoing tests in the 40 by 80 foot wind tunnel, May 1975.

Diverse Challenges Explored with Unified Spirit: 1969 – 1989

for the underlying silica tiles. In 1975 RCG was adopted for use over three-quarters of the orbiter surface. Ames also developed the LI-2200 tile, which was stronger and more refractory. This new tile, adopted in 1976, replaced one-tenth of the tiles on the orbiter Columbia.

When the 60 megawatt arc jet came on line, in March 1975, Ames could test full-scale tile panels in flows running thirty minutes, which is twice as long as the Shuttle reentry time. Ames performed most of the arc jet runs to certify the Shuttle thermal protection system, often running two shifts to fully simulate the Shuttle's lifetime of 100 flights. From this, Ames scientists gained new insight into the aerodynamic heating resulting from plasma flow over complex heat shields. When Shuttle designers grew concerned about hot gas flows between tiles, the Ames thermal protection branch devised a gap filler—a ceramic cloth impregnated with a silicone polymer. Once adopted in 1981, few Ames gap fillers have ever had to be replaced.

NASA also hoped to replace the white tiles that covered the top surface of the Shuttle orbiters (called LRSI for low-temperature reusable surface insulation) with a material that was cheaper, lighter, less fragile and easier to maintain. So Ames worked with Johns Manville to devise a flexible silica blanket insulation (called AFRSI for advanced, flexible, reusable surface insulation). Beginning in 1978, the AFRSI replaced most of the white tiles on the four later Shuttle orbiters. As the orbiters extended their operational lives, Ames researchers continued to invent and test improved reusable surface insulation tiles. Ames devised a new family of materials, which led to an even stronger and lower-

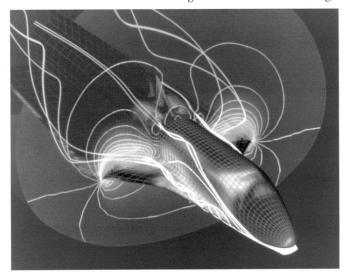

Computed image of flows around the Space Shuttle orbiter.

Artwork of the tiles underlining the shuttle Columbia show the subtle ghost-like patterns torched on their surface during reentry.

weight tile system (called FRCI-12 for fibrous refractory composite insulation) which was adopted in 1981 to replace one-tenth of the tile system. The insulation for the Shuttles has turned out to be lighter and easier to refurbish than previously expected, and has provided an excellent technical base on which to build the heat shielding for all future hypersonic vehicles.

Howard Goldstein, a leader of Ames' research on thermal protection systems.

Into the 1990s, led by Daniel Leiser and Daniel Rasky and guided by James Arnold, Ames continued to develop new thermal protection systems. David Stewart led Ames' basic research in catalycity—the study of how nitrogen and oxygen decompose in a shock wave then reform on a heat shield with lots of energy release—and made catalytic efficiency the basic measure for evaluating new insulators. An April 1994 mission with the shuttle Endeavour allowed the Ames thermal protection materials branch to test a new material (called TUFI for toughened uni-piece fibrous insulation) which is more resistant to impact damage from the dirt kicked up as the shuttle lands. Another new tile (called AETB for alumina enhanced thermal barrier) was adopted to replace tiles as the Shuttle further extends its operational life into the new century.

A new class of hypersonic vehicles and reusable launch vehicles under development in the late 1990s—such as the X-33, the X-34, the X-38 and the Kistler K-1—all depend upon Ames' work in thermal protection. Jeff Bull, Daniel Rasky and Paul Kolodziej of Ames also developed a

Diverse Challenges Explored with Unified Spirit: 1969 – 1989

Thermal protection materials developed at Ames for the Space Shuttle: AFRSI, GAP Fillers and FRCI-12.

very high temperature ceramic that will finally allow reentry vehicles to have a pointed leading edge rather than a blunt shape. In addition, Huy Tran led a team developing a silicon-ceramic heat shield for the Mars Pathfinder, and a phenolic-carbon ablating heat shield for the Star Dust asteroid return mission and the Mars sample return mission.

PLANETARY SCIENCE

The study of planetary atmospheres became a natural area of inquiry for Ames, since it merged work in the life sciences, atmosphere entry, aerodynamics and instrumentation with efficient project management.

During the Apollo years Ames had begun work in space science. Donald Gault had used Ames vertical gas gun to study cratering and meteoritics, information needed then for picking lunar landing spots. This information then grew in importance as scientists learned more about the role of impacts in the evolution of all planets. Charles Sonett led work on magnetometers, and John Wolfe, Vernon Rossow, and John Spreiter did work on solar plasmas. Carr Neel, John Dimeff and others in Ames' instrumentation branch built the sensors.

1 AFRSI

2 Gap Fillers

3 FRCI-12

Schlieren image of a straight-wing orbiter model being tested for stability and control characteristics in the 6 foot supersonic tunnel at Mach 95.

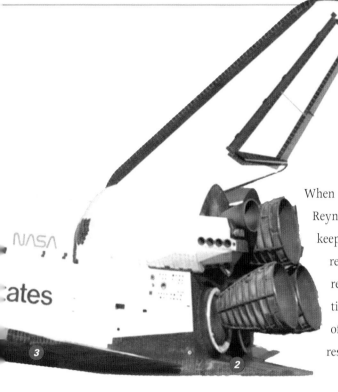

When better satellites travelled beyond the magnetosphere, Ray Reynolds led efforts to expand the Ames space sciences division to keep abreast of the data coming in. By the mid-1970s, a space science renaissance was born of the incredible diversity of data being returned—from the Pioneers to Jupiter and Saturn, Earth observation aircraft, the Viking landers, and the atmospheric probes. Years of planning and calibration culminated in a flurry of spectacular results from probes Ames had sent all over the solar system.

In the early 1960s, Alvin Seiff and David Reese began to explore the idea that a probe entering the atmosphere of a planet could determine the atmosphere's structure (density, pressure and temperature variation) as well as its composition. This idea emerged as Ames' vehicle environments division first considered the problems of landing a human mission on Mars through its still unknown atmosphere. Since the probe would enter at a very high speed, and perhaps burn up, it could carry no direct-measuring sensors. Accelerometers, instead, would measure deceleration in the air speeds which aerodynamicists used to compute atmospheric density and pressure. Temperature yielded information on the molecular weight of the atmosphere, so long as the aerodynamics of the probe were calibrated in the Ames tunnels over a variety of Mach and Reynolds numbers and in a variety of gases. The idea was intriguing to a great many aerodynamicists at Ames, who were accustomed to defining an atmosphere then

Bonnie Dalton reviews the Viking test module laboratory, March 1976.

Chuck Klein looking into the Mars Box, which simulated the environment of Mars for the preparation of life sciences experiments for the Viking lander.

Diverse Challenges Explored with Unified Spirit: 1969 – 1989

Atmosphere of Freedom Sixty Years at the NASA Ames Research Center

designing an aircraft configuration to produce the aerodynamic performance they wanted. Seiff turned the problem on its head—defining the configuration and performance to understand the atmosphere. Work began immediately in the hypersonic free flight facility, and with probe models dropped from aircraft.

The precursor to all of Ames' work in planetary probes was the June 1971 planetary atmosphere experiments test (PAET). PAET used what Ames had learned about reentry and hypersonics to push the frontiers of planetary studies. PAET was a complete prototype of the planetary probes to follow. It carried accelerometers, pressure and temperature sensors, two instruments to measure the composition of earth's atmosphere, a mass spectrometer and a shock layer radiometer. A Scout rocket launched from Wallops Island Station boosted the PAET out of Earth's atmosphere. A third stage rotated it back toward Earth, and a fourth rocket stage shot it into the atmosphere at 15,000 miles per hour. The data it returned validated the concept of the atmosphere entry probe—after scientists found an almost perfect match between PAET data and conventional meteorological data on atmospheric conditions. This provided the confidence to build probes to survey the atmospheres of other planets.

The 34 kilometer diameter impact crater, Golubkina, on the surface of Venus.

The probes for Pioneer Venus were not designed to withstand impact with the Venusian surface. But one probe did, and transmitted back data for 67 minutes. This is an artist's conception of what it might have looked like on the hot surface of Venus.

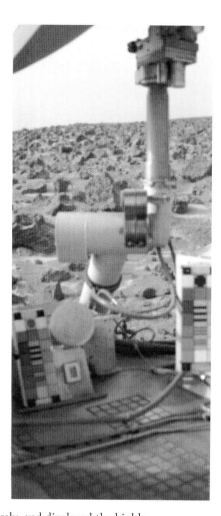

This photo from the Viking 2 lander shows white frost on the red Martian soil. By observing when it formed and when it melted, and matching that information with other data about the Martian atmosphere, Ames scientists were able to theorize about its chemical composition, October 1977.

The two Viking landers that settled down on the surface of Mars in September 1976 carried an atmosphere structure experiment designed by Seiff. Though not a probe, it provided the first detailed sounding of the structure of the Martian atmosphere. The Viking landers also included what would be Ames' first astrobiology experiment—a life detection experiment built by the Ames life sciences division, led by Chuck Klein. After Earth, Mars is the most likely planet in our solar system to support life. To search for such life, Vance Oyama built a gas exchange laboratory around a gas chromatograph to measure gas respiration in the Martian soil as it was treated with biological nutrients. It was a complex design: an arm extended to collect a sample, drop it in a jar, mix it with chemicals, and define the resultant gas. The gas exchange experiment worked flawlessly, and displayed the highly reactive chemical structure of the Martian soil. It found no evidence of life, though questions about what it did find motivated planetary scientists for years to come.

Vance Oyama at the gas chromatograph in Ames' life detection laboratory. Vance and his brother Jiro both pioneered new areas of life sciences research at Ames.

Pioneer Venus

The Pioneer Venus program was initiated in the same spirit as the earlier generations of Pioneer spacecraft—as a faster, better, and cheaper way of generating data about the atmosphere of Venus. It was managed by many of the same team, on the same management principles, with the same thirty month schedule, an equally conservative approach to engineering, and a simple set of "rules of the road for Pioneer Venus investigators" that kept the science paramount and focused. The mission to Venus earlier had been proposed to NASA by two atmosphere scientists—Richard Goody of Harvard University and

Donald Hunten of the University of Arizona. Based largely on the spectacular results of the PAET, NASA headquarters cancelled the Planetary Explorer program from Goddard, in January 1972, and opened in its place a Pioneer Venus group at Ames. Charles Hall led the group as Pioneer project manager, and Hughes Aircraft built the spacecraft. Among the experiments selected competitively to be included on the probes were those devised by four Ames researchers: Alvin Seiff on atmosphere structure, Vance Oyama on atmosphere composition, Boris Ragent on cloud detection and Robert Boese on radiative deposition.

The Pioneer Venus spacecraft had two components: an orbiter (Pioneer 12) that carried scientific instruments and a multiprobe bus (Pioneer 13) that launched the four probes into the atmosphere. The orbiter was launched on 20 May 1978; the multiprobe on 8 August. By 4 December the orbiter was in place and, five days later, the probes were dropped. Together, they returned data on the most thorough survey of another planet ever made.

Ames built each probe to known aerodynamic parameters so that its motion in flight, at an initial speed of 26,100 miles per hour, indicated the density of the atmosphere through which it travelled. As the probes

A paper collage interpreting the craters and ridged planes of Mars—and the Viking 2 as it passed over Mars' surface, on 2 November 1982, prior to landing.

The Pioneer Venus multiprobe bus depicted shortly after the probes had been released: (top to bottom) night probe, day probe, sounder probe, North probe.

heated up and interacted chemically with the atmosphere, they relayed data back to Earth on the climate, chemical makeup, and the complicated structure of the Venusian atmosphere. The Pioneer Venus science team found, for example, that there were remarkably small temperature differences below the clouds compared with the differences above, that the solar wind shapes Venus' ionosphere, and that the wavelike patterns visible from Earth are in fact strong wind patterns. They quantified the runaway greenhouse effect that makes the planet surface very hot. They identified widely varying wind speeds in the three major layers of clouds and a layer of smog, nine miles thick, atop the clouds. Using technology developed for the Viking gas exchange experiment, the Pioneer Venus orbiter first discovered the caustic nature of the Venusian atmosphere. They found that the surface was incredibly dry, and described the chemical process by which Venus' hydrogen blew off and its oxygen absorbed into surface rocks. They also measured its electrical activity, looking for evidence of lightning. Using these data and data returned from the Soviet Venera spacecraft, Ames scientists—James Pollack, James Kasting, and Tom Ackerman—proposed new theories of the origins of Venus' extreme atmosphere.

With the orbiter's precision radar, the Pioneer Venus team drew the first topographic maps of the cloud-enshrouded Venusian surface. They discovered that Venus had no magnetic field, from which they deduced that Venus had no solidifying core. They further discovered that Venus lacked the horizontal plate tectonics that dominated Earth's surface geology.

Early in 1986, Ames mission controllers reoriented Pioneer Venus, still in orbit around Venus, to observe Comet Halley. It was the only spacecraft in position to

Artist's concept of the Galileo probe separating from its heat shield.

observe the comet at its most spectacular—at perihelion, where it comes closest to the Sun and is most active. With Pioneer's ultraviolet spectrometer pointed at Halley, Ames scientists gathered data on the comet's gas composition, water vaporization rate, and gas-to-dust ratio. Five more times, mission controllers at Ames reoriented the Pioneer orbiter to observe passing comets.

The Pioneer Venus orbiter continued to circle the planet, working perfectly, for fourteen years—over one full cycle of solar activity. Its mission ended in October 1992, when controllers directed it into ever-closer orbits until it finally burned up. In doing so, it returned the best data yet supporting the theory that Venus was once very wet. For a cost averaging $5 million per year over its fourteen-year mission, Pioneer Venus generated a wealth of good science. By 1994, more than a thousand scientific papers had been written from Pioneer Venus data, authored by scientists from 34 universities, 14 federal laboratories, and 15 industrial laboratories. While planetary scientists continued mining Pioneer Venus data, the Ames people who built it turned their expertise to building similar probes for the atmospheres of Mars and Jupiter.

Galileo Jupiter Probe

Jupiter's atmosphere presented by far the biggest challenge for Ames planetary probe builders. Jupiter's huge gravity will accelerate a probe more than five times faster than the gravitational pull of the inner planets. Jupiter's enormous thermal and radiation energy and violent cloud layers are ominous spacecraft hazards. Jupiter has no recognizable surface; its deep atmosphere just gets denser and hotter until the edge blurs between atmosphere and any solid interior. Ames scientists expected any Jupiter probe to encounter 100 times the heat of an Apollo reentry capsule—something like a small nuclear explosion.

The Galileo descent module and heat shield prepared for launch.

Ames managed the Galileo probe project, and Hughes Aircraft of El Segundo built it. Robert Boese developed a net flux radiometer, Boris Ragent developed a nephelometer to measure the scatterings of cloud particles, James Pollack and David Atkinson devised a Doppler winds experiment, and Al Seiff led the probe atmosphere structure experiment—measuring pressure, temperature and density—culminating work he began in the late 1950s on the use of entry probes to define planetary atmospheres. Ames built a unique outer planets arc jet, led by Howard Stine and James Jedlicka, to simulate the most caustic and stressful atmosphere a man-made material would ever encounter. After computing and testing various exotic materials for their ability to withstand the heat, shocks, and spallation from the Jovian atmosphere, Ames chose carbon phenolic from which to engineer the massive heat shield needed to protect the probe as it entered Jupiter's atmosphere.

Hughes delivered the probe on schedule in February 1984, expecting an encounter in May 1988. Then it sat in storage for eight years. Galileo was designed to be launched from the bay of the Space Shuttle orbiter, but the Challenger accident threw the launch schedule into turmoil. In January 1988 NASA sent Galileo, now eight years old, back to Hughes for refurbishment and performance checks. Galileo was finally launched in October 1989, with a less powerful upper stage rocket and a more convoluted flight plan—one taking it by Venus and Earth to pick up speed on its journey toward Jupiter. Between design and launch, Benny Chin had taken over as probe project manager from Joel Sperans, Richard Young had taken over as project scientist from Larry Colin, and John Givens arrived as probe development manager.

Gus Guastaferro and Nick Vojvodich review their design of the Galileo heat shield.

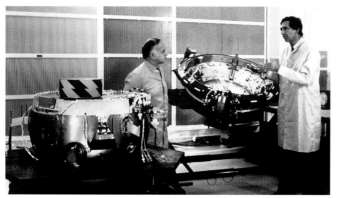

After travelling six years and 2.5 billion miles to Jupiter with the Galileo orbiter, the probe separated and entered Jupiter's atmosphere on 7 December 1995. The probe slammed into the atmosphere travelling 115,000 miles per hour, with deceleration forces 227 times Earth gravity. The incandescent gas cap ahead of the

Diverse Challenges Explored with Unified Spirit: 1969 – 1989

Heat shield of the Galileo Jupiter probe poised for a test run in the 12 foot pressure tunnel, June 1981.

heat shield reached 28,000 degrees Fahrenheit, meaning to an observer on Jupiter it glowed as bright as the Sun. Almost half of the probe mass was heat shield, most of which ablated away and the remainder of which fell away as the parachute deployed to slow its descent.

Seven instruments sent data back to the Galileo orbiter where it was stored for relay to the Jet Propulsion Laboratory. But soon after the encounter, the Galileo orbiter went over the horizon, then followed Jupiter behind the Sun, clouding the radio signal with noise. Scientists had to wait three long months for the complete return of data. Data received the following Spring confirmed that in the hour before it went dead under the pressure of the atmosphere, the Galileo probe returned the first direct measurements of the chemical composition and physical structure of Jupiter's clouds. The probe entered a hotspot—a gap in the clouds where the atmosphere was dry and deficient in ammonia and hydrogen sulfide. The probe survived to a depth of 22 atmospheres, sending data on atmospheric conditions and dynamics the whole way in.

Airborne Sciences

Meanwhile, Ames scientists studied Earth's atmosphere with equal fervor. Ames rebuilt its fleet of aircraft and outfitted them as flying laboratories used to conduct research in airborne science and Earth observation. Ames' medium-altitude aircraft included a

Painting depicting the Galileo spacecraft during vacuum tests at the Jet Propulsion Laboratory.

The Ames C-130 showing its bottom camera bays.

Photographs taken from the Ames C-130 show the 1988 Yellowstone fires in a composite of visible and thermal channels.

Learjet, a Convair 990 named Galileo II, and a Lockheed C-130.

The Learjet, though most often used for infrared astronomy, also proved useful in atmospheric studies of low-altitude wind shear in the 1970s. The Lockheed C-130 focused on Earth resources—in support of agriculture, meteorology and geology—and carried sophisticated equipment for mapping cropland, soils and nonrenewable resources. The C-130, equipped with a thermal infrared mapping sensor, was often called into service throughout the western United States to locate hot spots obscured by the dense smoke over forest fires. (And Ames researchers, ever interested in applying all their expertise to solving problems, in 1994 developed a low-cost electronic chart display to coordinate the many aircraft navigating around such large fires.) George Alger of Ames' medium-altitude missions branch led the C-130 in a variety of meteorology missions looking, for example, at biogeochemical cycling—how land interacts with the atmosphere.

Galileo II was the fastest aircraft in the fleet, and accommodated international teams of 35 researchers. This made it

Galileo I in the air in 1969.

especially valuable for global atmospheric research. Observers aboard Galileo II explored the origins of monsoons in India, interactions between ice, ocean and atmosphere off the northern coast of Greenland, and global atmospheric effects from the eruption of the Mexican volcano El Chicon. In 1990, Galileo II flew a research team led by Charles Duller that verified the discovery of a crater rim along the Yucatan peninsula. This provided evidence for a cometary or asteroid impact on Earth that might have led to the extinction of the dinosaurs.

Ames' first high-altitude aircraft, capable of flying to 70,000 feet, were two Lockheed U-2Cs that arrived in June 1971. As with so many research tools acquired during Mark's tenure as director, the U-2s were grabbed as surplus from another agency. The U.S. Air Force had announced that it would make the U-2s available for basic research. NASA was then in final preparations for the earth resources technology satellite (ERTS), managed by Goddard, and scientists were concerned that infrared and spectral-band photographs obtained on ERTS might be distorted because they would be taken through the entirety of Earth's atmosphere. The Air Force tasked Martin Knutson, one of the first U-2 pilots, to evaluate Ames' ability to fly and maintain the U-2s, which were notoriously slender and sensitive aircraft. Knutson then retired from the Air Force and joined Ames' airborne sciences office to lead the Earth Resources Aircraft Project to simulate the data collection process from the ERTS satellite. When delays meant the ERTS would miss its opportunity to survey chlorophyll levels in American crops during the 1972 summer growing season, Ames leapt to a plan and with three months of flights completed the entire benchmark survey with the U-2s. From there, research uses for the U-2s branched in many directions. In 1972, NASA headquarters designated Ames its lead center in Earth-observation aircraft

The ER-2 earth resources aircraft on the Ames tarmac surrounded by all the equipment that can be installed to image Earth.

and as a liaison to the scientific community. In response, Ames established an atmospheric experiments branch.

In June 1981, the U-2s were joined by a Lockheed ER-2 (for earth resources), a civilian version of the U-2. In May 1988 Ames acquired a second ER-2, and retired its thirty-year old U-2C. (Before being retired to static display at an Air Force base, this U-2C shattered sixteen world aviation records at Dryden for time-to-climb and altitude in horizontal flight, to 73,700 feet. These records were the first official acknowledgment of the U-2's previously classified altitude capability.) NASA and Lockheed Martin would later share a Collier trophy for development of the ER-2. Compared with the U-2, the ER-2

High-altitude ER-2 imagery of the Santa Clara Valley in 1989 showing Moffett Field and the surrounding wetlands in natural color.

Atmosphere of Freedom Sixty Years at the NASA Ames Research Center

Looking down into the cockpit of the NASA ER-2 aircraft, as Stanley Scott is preparing a meteorological experiment for the January 1989 airborne arctic stratospheric ozone expedition near Stavanger, Norway.

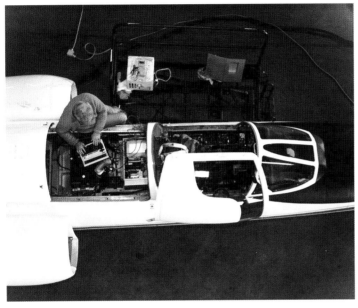

was thirty percent larger, carried twice the payload, had a range of 3,000 miles, had a flight duration of eight hours, and had four pressurized modular experiment compartments. In addition, Ames modified a DC-8 airliner into a flying laboratory for Earth and atmospheric sensing and for other key roles in NASA's Mission to Planet Earth. Ames often teamed the DC-8 and ER-2s on specific projects.

Ames scheduled the ER-2s flexibly enough, and built basing alliances with 42 airports around the world, so that Ames pilots could use them for quick-response storm observation, atmospheric sampling, and disaster assessment. The Ames U-2 measured ash cloud dispersement following the May 1980 eruption of Mount Saint Helens in Washington state. Life scientists at Ames and the University of California at Davis used remote-sensing data on vegetation growth, collected between 1984 and 1988, to devise a model that actually predicted the spread of mosquitos that carried malaria. Similar remote spectral scanners were used in April 1993 for Project GRAPES, an effort to plot the spread of phylloxera infestation through California vineyards. The ER-2s proved especially useful in calibrating new remote-sensing equipment flown aboard LANDSAT Earth-observation satellites and the Space Shuttle. In 1989 and 1990, the DC-8 flew the global backscatter experiment (GLOBE) to survey airborne aerosols in the Pacific basin and test out new experiment packages designed for the Earth Observing System satellite. In February 1993, Rudolf Pueschel and Francisco Valero of the Ames

The interior of the Galileo II in 1972. Ames used this converted Convair 990 as an airborne science platform.

ER-2 on Ames Ramp with pilot James Barrillearx entering the cockpit.

atmospheric physics branch led the DC-8 and an ER-2 to Australia to map the interior of a tropical cyclone and explore the coupling of the atmosphere and the warm ocean.

Perhaps the most significant research done by Ames' airborne scientists was the many-year exploration of Earth's ozone layer. In August and September 1987, operating from Punta Arenas at the southern tip of Chile, Ames scientists used the ER-2 and the DC-8 to make the first measurements that implicated human-made materials in the destruction of stratospheric ozone over Antarctica. During the winter of 1989, the ER-2 and DC-8 team, led by Estelle Condon and Brian Toon and based in Norway, completed an airborne campaign to study ozone chemistry and distribution over the Arctic. The ER-2 and DC-8 returned to the Arctic in 1992 to map changes in stratospheric ozone, and the results of their work were written into the Montreal Accord on limiting chemicals that deplete the ozone.

Infrared Astronomy

The other airborne platforms in Ames' fleet played a key role in the growth of the discipline of infrared astronomy. Until the 1960s, the main reason telescopes were mounted on airplanes was to follow solar eclipses. But the invention, in 1961, of a germanium bolometer able to detect infrared radiation up to 1,000 microns in wavelength opened up the age of infrared astronomy.

The ancients gazed into the night sky and saw a majestic canopy of changeless stars. Optical telescopes and spectrographs of great power further unveiled the immensity

The 91 centimeter airborne infrared telescope model.

Diverse Challenges Explored with Unified Spirit: 1969 – 1989

and complexity of the universe but always within a small window—wavelengths that were both visible and that made their way through Earth's atmosphere. Aircraft, then spacecraft, let astronomers place their instruments far above the obscuring water vapor of the

atmosphere where they could see all the messages that the universe was sending us—all the radiation, from all the sources, at all the wavelengths. Infrared (or heat) radiation conveys information about the composition and structure of

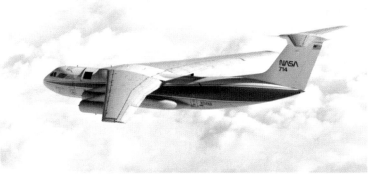

The Kuiper Airborne Observatory, with telescope door opened.

Earth-bound solids and gases. It also penetrates the dense clouds of dust that obscure regions where stars and planets are forming. Infrared observation became our best source of information about the chemical composition of remote planets, stars and nebulae.

Ames started its work in infrared astronomy in 1964, soon after Michael Bader, chief of the Ames physics branch, returned from a very successful airborne expedition to observe a solar eclipse. Ames purchased an old Convair 990 aircraft, named it Galileo and began converting it into an airborne science platform. Along the upper left side of the fuselage, Ames mechanics installed thirteen 12 inch apertures of optical-quality glass in time for the solar eclipse of 30 May 1965. From the beginning, Ames made its airborne science expeditions open to scientists from around the world. They made observations of three solar eclipses, the comet Ikeya-Seki, Mars during opposition, and the Giacobini meteor shower. Using a telescope with a gyrostabilized heliostat for precise pointing, one team of scientists obtained a remarkable set of near-infrared spectra for Venus, showing that the Venusian clouds were not

View inside the telescope door of the Kuiper Airborne Observatory.

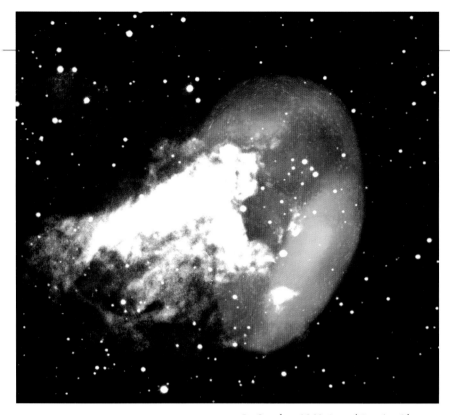

A composite image, taken aboard the Kuiper Airborne Observatory, of Messier object M 17, known as the Omega Nebula.

made of water as suspected. Later flights showed they were made of sulphuric acid droplets. In 1973, the Galileo was tragically lost in a mid-air collision with a Navy P-3 near Moffett Field that killed everyone on board. It was replaced by another Convair, named Galileo II, though it was used primarily for Earth observation.

In October 1968 Ames' Learjet Observatory made its first observations. Its apertures were larger than those on the Galileo and opened to the sky without an infrared-blocking quartz cover. Flying above 50,000 feet, teams of two observers aboard the Learjet discovered a host of bright infrared sources. They measured the internal energies of Jupiter and Saturn, made far-infrared observations of the Orion nebula, studied star formation regions, measured water in the Martian atmosphere, and generally pioneered astronomy in the wavelength range of 30 to 300 microns. Ames also used the Learjet to observe events around Earth, like eclipses and occultations.

Encouraged by the success of the Learjet, Ames built the much larger Kuiper Airborne Observatory (KAO). The KAO platform was a military transport aircraft (a Lockheed C-141 Starlifter) housing a 36 inch reflecting telescope in an open port. Soon after its first observations in January 1974, it was renamed in honor of Gerald P. Kuiper, director of the Lunar and Planetary Laboratory at the University of Arizona and a leading light in infrared astronomy. The KAO flew only as high as 45,000 feet, yet was a big advance over the Learjet. It accommodated up to twenty scientists, flew missions over 7.5 hours

Steve Butow and Mike Koop aboard the KAO set up equipment to observe a Leonid meteor shower.

Diverse Challenges Explored with Unified Spirit: 1969 – 1989

The Infrared Astronomy Satellite (IRAS), November 1983.

long, and averaged seventy missions per year. Most importantly, the KAO telescope was balanced on a 16 inch diameter spherical air bearing (the largest ever constructed) and was completely gyrostabilized so it would not be bounced around by air turbulence. Light from the telescope passed through the air bearing and into the variety of instruments attended by scientists in the pressurized cabin.

Observers on the KAO made many significant discoveries: they found the rings around Uranus; mapped a heat source within Neptune; discovered Pluto's atmosphere; detected water vapor in comets; explored the structure and chemical composition of Supernova 1987a; mapped the luminosity, dust, and gas distributions at the Milky Way's galactic center; and described the structure of star-forming clouds. Jesse Bregman developed a spectrograph used with the KAO telescope that in June 1993 detected water molecules on the surface of Jupiter's moon Io. (Laboratory work in 1988 on planetary ices by Farid Salama first suggested the presence of water on Io.) They also discovered 63 spectral features—atomic, molecular, solid state—of interstellar materials. Before the KAO, astronomers had identified only five molecular species. KAO observers identified 35 others throughout the galaxy. As important as all these scientific breakthroughs was that a generation of infrared astronomers were trained on the KAO.

Ames researchers applied their expertise in airborne observatories to the design of spaceborne observatories. Ames worked with scientists at the Jet Propulsion Laboratory, in the Netherlands and the United Kingdom to design the complete Infrared Astronomy Satellite (IRAS). Ames itself created the IRAS telescope, which has a 60 centimeter mirror and

Airborne telescope and its control console being prepared at Ames for installation in the Lockheed C-141 Starlifter aircraft that served as the Kuiper Airborne Observatory.

The power of infrared astronomy is displayed in these 1983 photographs. The dark lines in the bottom photo, taken of the Milky Way in visible light, are clouds of dust that obscure our view of the stars behind them. The real shape of our galaxy is revealed in the infrared image (top) obtained by the infrared astronomy satellite. Infrared light penetrates the dust clouds to show that the galaxy appears as a thin disk, just like the edge-on spiral galaxies we see throughout the cosmos.

an array of detectors cooled to near absolute zero by superfluid helium. It was launched in January 1983 and, during the one year it survived in orbit, IRAS made the first whole-sky survey ever conducted in the infrared region. In mapping the entire celestial sphere in four infrared bands from 8 to 120 micrometers, IRAS astronomers found 250,000 new infrared sources, suggestions of asteroidal collisions in the zodiacal cloud, particle rings around some stars, and the cool, wispy filaments of the infrared cirrus covering much of the sky. And IRAS returned valuable experience useful in building the next generations of airborne telescopes.

With its infrared astronomy and planetary probes, Ames scientists gathered huge data sets on the molecular dynamics of the universe and on the chemical composition of our solar system. With the airborne science experiments, Ames was calibrating that data with all that we knew about Earth. Ames people wanted to make sure that those hard-won data were well used and, in sorting through every nuance, they made extraordinary advances in planetary science.

Exobiology, Astrochemistry and the Origins of Planetary Systems

Exobiology continued to be a major focus at Ames, though tied ever more closely to Ames' work in space science. Sherwood Chang led the planetary biology branch and, along with Ted Bunch, did pathbreaking work on organic material and water in meteorites. David Des Marais and Christopher McKay studied the intricate lives of some of Earth's most primitive microorganisms, while Jack Farmer, David Blake, and Linda Jahnke studied the fossil markers for extinct microbial life. This led to a series of bold explorations to find organisms in extreme environments—hot springs, Antarctic deserts, and frozen lakes. Finding organisms in those places was good practice, they thought, for finding life on Mars. Exobiology may have been the science without a subject matter, but Ames indeed found good proxies.

Donald DeVincenzi, the exobiology program manager at NASA headquarters, supported Ames efforts to host workshops and write the papers that continued to define the scientific core of the discipline. A July 1988 meeting with the International Astronomical Union addressed the chemical composition of interstellar dust. Others presented pathbreaking work on the presence of carbon in the galaxy. As NASA missions returned new data on solar system bodies—Venus, Mars, asteroids, comets, Europa and the gas planets—Ames exobiologists studied them for clues to the possibilities of life. Similarly, when new missions were planned—like Titan-Cassini or the Mars rover sample return—Ames exobiologists made sure that the biological experiments were well conceived.

Ray Reynolds had done theoretical space science on the formation of planets at Ames since 1964, well before Ames had begun managing any of its space or observational missions. Hans Mark, like the American public, was fascinated by planetary exploration and supported Reynolds' efforts to build a world-class theoretical studies branch in space science. David Black, who first discovered signs of interstellar material in a meteorite, came to Ames and built the Center for Star Formation Studies. The Center was a consortium of Ames and two University of California astronomy departments (at Berkeley and Santa Cruz) and greatly advanced the astrophysical theory of protostellar collapse. They used supercomputers well: they modeled systems ruled by self-gravitation, like galaxies, protostellar clouds, and solar nebula; ran three-dimensional, n-body calculations that followed the motions of billions of stars in their own gravitational fields; calculated the collapse of rotating interstellar clouds to ten orders of magnitude in density; demonstrated that the true shape of elliptical galaxies was prolate rather than oblate; and showed

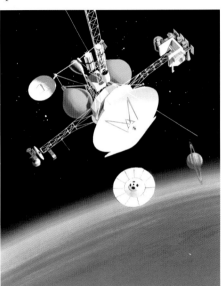

Saturn orbiter and Titan probe spacecraft.

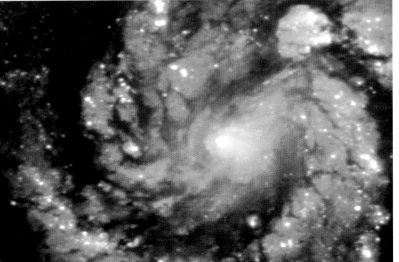

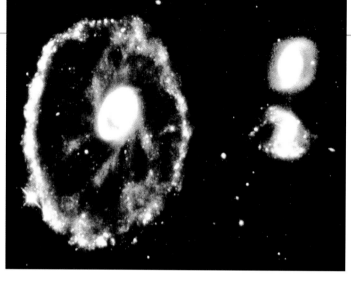

A rare and spectacular head-on collision between two galaxies appears in this NASA Hubble Space Telescope true-color image of the Cartwheel Galaxy, located 500 million light-years away.

how galaxies collided. Reynolds also hired Jim Pollack.

James Pollack, a radiative transfer theorist in the planetary systems branch of the Ames' space sciences division, arrived at Ames in 1970. He always seemed to come up with ingenious ways of connecting some theoretical insight, with the tools Ames had available, and with the scientific challenges people were wrestling with. In the 24 years Pollack worked at Ames before his death, he wrote nearly 300 articles on all facets of planetary science. Postdoctoral fellowships offered by the National Research Council fed much of the scientific vigor at Ames, especially in the planetary sciences. The best young scientists came to Ames for two-year projects, often to work with Pollack, and the best of those hired on. A great many others came to hang experiments on NASA spacecraft or to mine NASA data.

James Pollack developed models of planetary atmospheres and planetary evolution and verified them in laboratory, airborne and spaceborne experiments.

Pollack's drive to understand the origins of planets and the evolution of their atmospheres—especially for the "habitable" planets like Earth, Mars and early Venus—led him to use any variety of numerical, observational, or experimental tools. Pollack worked with Richard Young and Robert Haberle to develop an entire suite of numerical models of the climate and meteorology of Mars. These models comprised a unique resource—used to plan Mars missions, analyze the data they returned, and advance theories on how the climate of Mars changed over eons as the Sun warmed up and Mars' atmosphere escaped. The Ames team devised similar numerical models to explain the greenhouse gas climate of Venus, its high surface heat, its current lack of water, and its acidic atmosphere. Pollack inevitably teamed with other environmentally concerned researchers exploring the atmosphere of Earth. With James Kasting and Thomas Ackerman, he initiated some of the first studies of atmospheric aerosols and their effect on the evolution of Earth's climate. Brian Toon contributed his expertise on cloud microphysics, thus bridging efforts in the planetary sciences and Ames' Earth-observation aircraft. These colleagues led the team that later wrote the famous paper on "nuclear winter," suggesting that dust

Atmosphere of Freedom

Remote sensing discovered ancient impact craters believed to result from the impact that scientists see as the key to the dinosaurs' disappearance.

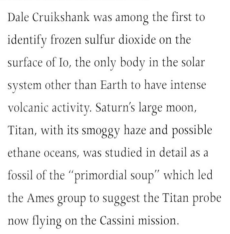

and soot kicked into the atmosphere by a nuclear war would degrade the habitability of Earth as much as the comet impacts that reshaped the climates of other planets and that might have led to the demise of the dinosaurs.

Voyager's grand tour of the outer solar system, coupled with data returned from the Pioneers and observatories, drove a revolution in planetary science focused on the evolution of Jupiter, Saturn and their moons. Pollack, Reynolds and their collaborators wrote stellar evolution codes to explain the residual internal heat of these gas giants, their growth by accumulation of planetesimals, and the subsequent capture of hydrogen envelopes. Jeff Cuzzi, Jack Lissauer and their collaborators unravelled puzzles in the rings of Saturn and the other gas giants, including spiral waves, embedded moonlets, and their rapid evolution under meteoroid bombardment.

Dale Cruikshank was among the first to identify frozen sulfur dioxide on the surface of Io, the only body in the solar system other than Earth to have intense volcanic activity. Saturn's large moon, Titan, with its smoggy haze and possible ethane oceans, was studied in detail as a fossil of the "primordial soup" which led the Ames group to suggest the Titan probe now flying on the Cassini mission.

Ames has also fueled interest in the origin of other planetary systems. Black led the first early studies techniques to find planets around other stars, which presaged future NASA planetary detection missions like Kepler. In addition, the Ames planetary scientists did pioneering studies of the gravitational and fluid dynamics of protoplanetary disks. Later, they connected the disciplines of astrophysics and meteoritics in studying planetary formation, often by leveraging Ames' in-house

Rings of Saturn.

Io, Jupiter's innermost moon.

expertise in aeronautical fluid dynamics.

Life is made from organic material. Into the early 1990s a unifying theme among Ames researchers was to chart the path of organic material from its origin in the interstellar medium (where infrared astronomy revealed it was formed), through primitive meteorites (available for chemical analysis), and into Earth's biosphere. David Hollenbach and Xander Tielens studied the physical evolution of grains in space. Lou Allamandola picked up the critical question of the chemical evolution of organic materials. It took him many years to piece together laboratory equipment to mimic the space environment and show how organic material could be produced from hydrogen, oxygen, carbon and nitrogen formed first in the big bang and then subsequently in stars. Allamandola's group showed how polycyclic aromatic hydrocarbons evolved from elementary carbon, and dominate infrared emissions from the Milky Way.

The unique atmosphere at Ames allowed all this work to cross-pollinate—in planetary formation, the evolution of planetary atmospheres, and the chemical, thermal and gravitational evolution of the solar system. It also coupled Ames' early pioneering work in

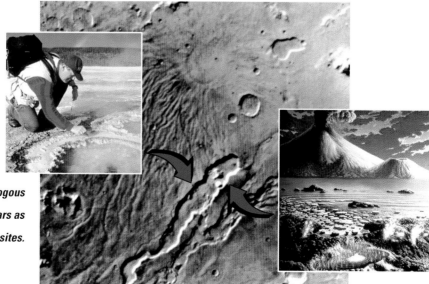

Primitive microorganisms thrive in hot springs on Earth, so Ames is identifying analogous ancient environments on Mars as potential landing sites.

Diverse Challenges Explored with Unified Spirit: 1969 – 1989

Barney Oliver was an early advocate of SETI, and guided its advances in signal processing.

exobiology and the chemical origins of life with the broader discipline later called astrobiology.

SETI (SEARCH FOR EXTRATERRESTRIAL INTELLIGENCE)

In the late 1960s, John Billingham of Ames' biotechnology branch, began to move Ames into the search for extraterrestrial intelligence (SETI). SETI seemed a natural area of interest for Ames. It combined the exobiology quest for life beyond Earth with space science theory for deciding where to look for it, and radio astronomy and computation as the means to search for it. In 1971, Billingham teamed with Bernard Oliver, a former vice president for research at the Hewlett Packard Company and a technical expert in microwave signal processing. They proposed Project Cyclops—$10 billion for a circular array of 1,000 telescope dishes, 100 meters in diameter, to do a full-sky survey of coherent microwave signals. But neither NASA headquarters nor its scientific advisors would endorse so expensive an effort in such uncertain science.

Jill Tarter provides the scientific vision for the SETI Institute.

Billingham also sketched more modest steps that NASA could take to help the many university astronomers engaged in SETI. Collectively, they decided to start searching for nonrandom radio waves in the microwave portion of the spectrum (microwaves travelled well in space and earthlings were already propagating them around the universe). They also decided to search between the natural spectral emission of hydrogen and the hydroxyl radical (OH)—dubbed the water hole—since water is essential for life.

Hans Mark began to appreciate the value of a comprehensive SETI program, not only for what it might discover, but also for what it could teach us about pulses in the universe and as a way to excite children about science. In July 1975, Mark asked NASA headquarters to fund a second international SETI meeting. Administrator James Fletcher instead obliged Mark to find money from the

John Billingham, John Wolfe, and Barney Oliver lead a 1976 discussion on the best strategies for searching for extraterrestrial intelligence.

National Academy of Sciences, but to hold the meeting at Ames. Fletcher did not want NASA to fund SETI prior to a formal commitment authorized by Congress. Over the next five years, and with Sy Syvertson's encouragement, Ames and JPL (which ran NASA's Deep Space Network) contributed a total of $1.5 million to design signal processing hardware and algorithms and to hold a series of workshops to map out the most appropriate scientific strategy for SETI. Billingham organized the series of multidisciplinary workshops that brought together a range of scholars—from astronomy, electronics, biology, psychology and philosophy—to debate the once taboo subject of contacting life beyond our solar system. Two regular attendees were Frank Drake and Philip Morrison, the first astronomers to lend credence to the subject by calculating the probabilities of extraterrestrial intelligence.

NASA began to fund SETI more seriously in 1981—at an average of $1.9 million per year over the next decade—but its value was constantly challenged. Senator William Proxmire had bestowed a Golden Fleece on the SETI program in 1978, and in 1981 Proxmire successfully passed an amendment deleting SETI's fiscal 1982 funding. Carl Sagan met with Proxmire to argue the merits of the science, and Proxmire agreed to no longer oppose the program. SETI backers became more politically active. They founded the nonprofit SETI Institute near Moffett Field, encouraged university astronomers to turn their ears skyward for highly focused searches, and got Soviet scientists to release data on their efforts. The FAA showed an interest in using frequency analyzers developed for SETI, and the National Security Agency learned about code breaking. SETI was small, well-managed, on budget, and returning interesting science—if not yet evidence of intelligent life, at least far better knowledge about the energy patterns in the universe.

Carl Sagan and David Morrison at the First International Conference on Circumstellar Habitable Zones.

On the 500th anniversary of Columbus' voyage to America, NASA formally launched a SETI program.

Diverse Challenges Explored with Unified Spirit: 1969 – 1989

Frog environment unit mock-up, prior to Spacelab J.

Renamed the high resolution microwave survey, it was funded by the NASA headquarters exobiology program, located at Ames and managed by project scientist Jill Tarter of the SETI Institute. It received $12 million in fiscal 1992 against a $100 million budget over ten years. After two decades of arguing over the mathematical probabilities of other intelligent life, Ames researchers finally got a chance to actually look for it in a systematic way. While scientists at JPL geared up for a lower-resolution sky survey of the full celestial sphere, Ames developed the equipment and algorithms for a targeted search of solar-type stars. Devices built at Ames would resolve 10 megahertz of spectrum into 10 million channels, simultaneously and in real time. The resulting coverage would have 100,000 times more bandwidth than devices used in previous searches, and was a billion times more comprehensive.

Yet less than a year later, Congress killed NASA's SETI/HRMS program. It died from fervor over the federal deficit and a history of unfounded associations with UFO encounters. The scientific community did not lobby consistently for it—SETI was an exobiology effort that used the tools of radio astronomy. To make it politically palatable, NASA had moved SETI from its life sciences to its space sciences directorate, which gave it low priority. Most damaging, NASA headquarters did not fight very hard to keep SETI in NASA's budget. SETI was small enough to sacrifice easily, and headquarters already felt bloodied from its 1992 budget encounter with Congress. The SETI Institute continues its work with private funding.

COSMOS/BION

A superb example of Ames' ability to do pioneering science quietly and on a small budget was the Cosmos/Bion missions. Every two to four years, between 1975 and 1997, the Soviets shot a Cosmos biosatellite into space carrying an

Twelve foot linear sled installed in Ames' vestibular research facility, 1987.

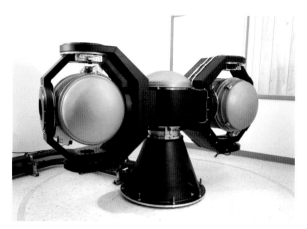

Vestibular research facility, opened in 1985 for studies of animals' inner ears.

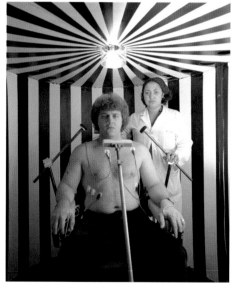

Patricia Cowings sets up a motion sickness study on Bill Toscano, who is preparing to ride the 20 g centrifuge in a specially designed cab.

array of Ames life science experiments to study the adaptability of plants and animals to microgravity. A unique spirit of cooperation underlay the success of Cosmos/Bion. Even in the darkest days of the Cold War—following the Soviet invasion of Afghanistan and the Reagan presidency—life scientists from Ames, western and eastern Europe, and the Institute for Biomedical Problems in Moscow continued to collaborate on basic research.

The Soviets had already flown two Cosmos biosatellites before inviting NASA to join the third, to be launched on 25 November 1975. Ames scientists jumped at the chance. The Ames Biosatellite program was cancelled in 1969, the promise of Skylab faded in 1973 as power failures crippled it, and the first biological payload on the Shuttle would not fly until 1983. While Ames had a superb set of ground-based centrifuges for use in studying the biological effects of hypergravity, the only way to study microgravity was in space. In addition, the Soviets offered to pay the entire cost of the spacecraft and launch; NASA need only pay for design and construction of experiment payloads to fly on board. During the 1970s, this never cost NASA more than $1 million per launch. For this relatively small cost, Ames produced some superb data.

The first launch, Cosmos 782, landed 19.5 days later in central Asia. For security reasons, Soviet scientists recovered the experiments and returned the samples to Moscow. The rat studies exemplified the success of the mission. Eighteen institutions from five countries did studies on every major physiological system in the rat. Many of these experiments were designed by people at Ames: Delbert Philpott of the Ames electron microscope laboratory studied radiation bombardment to the retina; Emily Holton measured bone density and renewal; Joan Vernikos studied gastric ulceration; Adrian Mandel evaluated immunity levels; Henry Leon measured degradation of red blood cells; and Stanley Ellis and Richard Grindeland charted hormonal levels. As experimental controls,

A Soviet Vostok biosatellite like those used in all the Cosmos/Bion missions.

the Soviets built a biosatellite mockup that stayed on the ground simulating every flight condition but weightlessness, as well as a small centrifuge for the biosatellite that kept a small control colony at 1 g of artificial gravity. Ames scientists concluded that the stress on the rats came from weightlessness rather than from other flight factors, that spaceflights up to three weeks generally were safe, but that specific results needed to be verified.

After the second flight, Cosmos 936 in August 1977, the results were clearer. Basic physiological systems showed no catastrophic damage, but there was measurable bone loss and muscle atrophy from exposure to microgravity, as well as retinal damage from radiation bombardment. Indeed, the regularity to the Cosmos/Bion flights let Ames biologists constantly improve their protocols and confirm their data. Ames scientists were initially unaccustomed to sending up experiment packages every two years, but they eagerly adapted to the quickened pace of data analysis, publication, experiment proposal, and payload design. New collaborators were added constantly, using new types of organisms—plants, tissue culture, fruit flies and fish. Every flight used a mass-produced spherical Vostok spacecraft—eight feet in diameter, a volume of 140 cubic feet, with active environmental control, and a payload of 2,000 pounds. Ames project engineer Robert Mah built the cages and bioinstrumentation to fit the space allocated by the Soviets.

Jiro Oyama in 1968 controlling a life sciences experiment in the Ames 50 foot centrifuge.

Frogs in space: Spacelab J flight frogs.

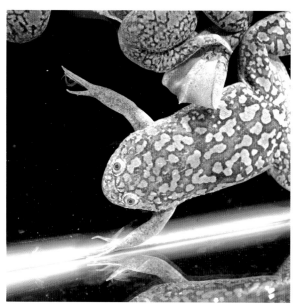

Sid Sun testing the glovebox of the centrifuge facility mock-up.

Kenneth Souza at Ames and Lawrence Chambers at NASA headquarters oversaw the entire program in one capacity or another, and the Soviets no doubt appreciated this continuity of leadership that was so rare within NASA. Eugene A. Ilyin led all efforts in Moscow, and Galina Tverskaya translated with graciousness and precision.

During the 1980s, the cost to NASA rose to an average of $2 million for each Cosmos/Bion mission, primarily because the mission added two rhesus monkeys as research subjects. The Soviets had never flown monkeys in space; the last time Americans tried, in 1969, the monkey died. So the Cosmos 1514 mission in December 1983 lasted only five days. Not until Cosmos 2044 in September 1989 would the monkeys fly a full two weeks. These flights displayed the remarkable progress Ames had made in bioinstrumentation. Specimens in the earliest Cosmos/Bion flights were flown undisturbed, and descriptive data were collected post-flight. For the later flights, the animal and plant specimens were fully instrumented and data was collected continuously during flight. James Connolly became project manager in 1985, and more consciously focused the Cosmos payload to complement those flown aboard the Shuttle.

The final Cosmos/Bion mission included a rhesus monkey project devised jointly by American and French scientists. It was originally designed to fly aboard the Shuttle, but was cancelled because of cost and sensitivity concerns. Ames had developed a well-established protocol for the low-cost development of biological experimentation, and quickly modified the rhesus project to fly on Bion 11 for $15 million, a fraction of the original cost. It launched on 24 December 1996 and landed fourteen days later with the monkeys in good health. However, a day later, during a biopsy procedure requiring anesthesia, one of the monkeys died. A panel of experts convened by NASA headquarters confirmed the validity and safety of the rhesus research. But animal rights activists vilified

A spin-off of Ames research on both bone density in microgravity and on thermal protection foams is this bone-growth implant shown, in 1993, by Daniel Leiser and Howard Goldstein in Ames' shuttle tile laboratory.

this death, and Congress questioned why NASA was spending money to help the Russians send monkeys into space. Indeed, with the dissolution of the Soviet Union, the Russians had begun asking NASA to fund a greater portion of the flights. Early in 1997 Congress refused to appropriate $15 million for the Cosmos/Bion mission planned for the summer of 1998. Few at Ames participated full time in Cosmos/Bion, so its cancellation had little impact on staffing levels. The cancellation, however, immediately degraded Ames efforts to pursue a systematic research program. The Cosmos/Bion program remained, as it will for the foreseeable future, the single best source of data on the effects of weightlessness on earthly life.

Gravitational Biology and Ecology

The Cosmos/Bion program was the free-flier portion of a much broader effort at Ames to explore the prospects of earthly life living in space—a program that also included Shuttle-flown and Earth-based experiments. On Earth, Ames continued to explore how humans responded to weightlessness. Dolores "Dee" O'Hara managed Ames' human research facility where, since the early 1960s, a great many Ames life scientists had refined bed rest into a superb tool for understanding specific responses to weightlessness. Bed rest with a head-down tilt of six degrees, for example, simulates the decreased blood volume incurred during space travel. Joan Vernikos, chief of Ames' life sciences division, used the bed-rest facility to determine which means of plasma expansion made fainting less likely upon return to Earth. She also studied how much gravity was required to remain healthy, supporting NASA's decision

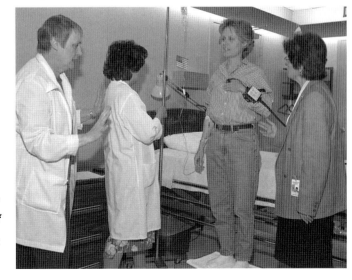

Dee O'Hara (at left) and Joan Vernikos (at right), in May 1993, prepare a volunteer for a bed rest study to simulate the effects of microgravity on the human body.

An experiment package on the circadian rhythms of the increment beetle being prepared for launch aboard the space shuttle to rendezvous with the Mir space station.

to provide intermittent gravity with an onboard centrifuge rather than rotating an entire space station. David Tomko directed the Ames vestibular research facility to coordinate the work of many Ames life scientists studying the body's system of balance and spatial orientation. Likewise, researchers interested in hypergravity worked closely with Robert Welch in the 20 g centrifuge, NASA's only human-rated centrifuge.

Spacelab, flown aboard the shuttle orbiter, carried Ames' experiment payloads in the early 1990s. The Ames space life sciences payloads office provided half of the experiments flown aboard the Spacelab Life Sciences-1 (SLS-1) mission in June 1991. As the first Spacelab mission dedicated to the life sciences, SLS-1 provided an opportunity to study the effects of weightlessness in a comprehensive fashion. The crew hooked on biomedical sensors, many developed at Ames, to study the effects of weightlessness, and ran experiments on animals and plants in the Ames payload. Bonnie Dalton was project manager and oversaw training of the mission specialist crew, coordination of the experiments, and development of new biosensors. The Ames payload included the research animal holding facility—providing life support to nineteen rats—and the general purpose work station—a glove box to contain liquids during experiments. Because this hardware tested perfectly, Ames could plan on in-flight animal testing in forthcoming missions.

In September 1992, two experiments from Ames investigators flew aboard the STS-47 Spacelab mission. Kenneth Souza designed a frog embryo experiment, Greg Schmidt served as payload manager, and Jack Connolly designed the "frog box." Not only was this the first time live frogs flew in space, but they would also shed eggs that would be fertilized and incubated in microgravity. The experiment showed that reproduction and maturation

Glovebox to contain the frog embryology experiment, being tested prior to being carried aboard the 1992 STS-47 Spacelab J Mission.

Atmosphere of Freedom Sixty Years at the NASA Ames Research Center

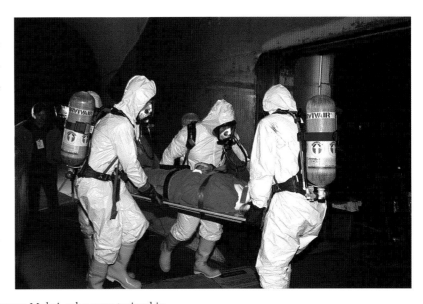

The Ames DART team provides disaster assistance and rescue wherever and whenever needed. In December 1998, in cooperation with a variety of local, state and federal agencies, they practiced their response to a collapsed building emergency.

can occur normally in space—at least with amphibian eggs. Biologists had studied amphibian eggs for more than a century because of the unique way they orient themselves to gravity once fertilized. Patricia Cowings, in an updated version of an experiment flown on Spacelab 3 in 1985, demonstrated that astronauts Mae Jemison and Momuro Mohri, who were trained in autogenic feedback, could alleviate symptoms of space motion sickness without medications using a "bio-belt" monitoring system built by Ames technicians.

SLS-2 (Spacelab Life Sciences-2) flew aboard the shuttle orbiter in October 1993, marking the first time ever that astronauts had collected tissues in space. Before then, all tissues were collected by the principal investigators after the flight landed, making it impossible to separate the physiological effects of microgravity from the hypergravity of liftoff and landing. Furthermore, the shuttle payload specialists first collected tissues on the second day in space—sacrificing five rats, doing rough dissections, and preserving the tissues—allowing life scientists back at Ames to do the fine dissections and to note how quickly the organisms adapted to space. Tissues were collected again on day fourteen, the day before reentry, so that life scientists could study how quickly the organisms readapted to Earth's gravity. The speeds of adaptation and readaptation were especially notable in experiments on bone density and neurological development. Martin Fettman, a veterinarian, flew as the payload specialist responsible for the rats, and Tad Savage and William Hines of Ames managed the payload of nine experiments.

To better apply to NASA missions all that Ames had learned about the adaptability of various organisms to microgravity, in March 1990 Ames created an advanced life support division. Initially led by William E. Berry and deputy Lynn Harper, the division developed bioregenerative and closed loop life support systems that would allow astronauts to

John Hines leads the Sensors 2000! program, which leverages Ames' rich tradition in instrumentation and life sciences to create ever smaller and more precise biomedical instruments.

William Mersman and Marcie Chartz (Smith), in October 1958, operating the new IBM 704 electronic computer.

colonize the Moon or travel for long periods to distant planets. Some systems were simple—like a self-contained salad machine designed by Robert MacElroy and Mark Kliss, to grow fresh vegetables aboard the space station. Some were very complex, like chemical and biological technologies to close the life support loop and enable nearly self-sufficient human habitats in space or on other planets. In addition, Bruce Webbon led efforts to design advanced spacesuit technologies for extravehicular activity and planetary exploration. Likewise, Ames consolidated its work in biotelemetry into a sensors development program, led by John Hines and later renamed the Sensor 2000! program, which developed new technologies for prenatal care in the womb. Throughout its sixty-year history, Ames' instrument builders—in both the life and physical sciences—have been key contributors to spin-off technologies for American industry. Another major contribution was computational fluid dynamics, built on the computing infrastructure at Ames.

COMPUTING AT AMES

Computational fluid dynamics (CFD)—using computers to depict airflows—was one of NASA's most important contributions to the American aerospace industry. CFD emerged as a scientific discipline largely because of work done at Ames. Two events mark its birth. Harvard Lomax, a theoretical aerodynamicist, in 1969 formed a computational fluid dynamics branch and recruited a world-class group of researchers to staff it. Second, in 1970, Ames negotiated the acquisition of the Illiac IV, the world's first parallel computer. As with most things at Ames, though, these two birthing events merely accelerated an established tradition.

Computers at Ames initially were women, hired to generate smooth curves from the raw data of tunnel and flight tests using electromechanical

Composite layout of all the pieces of the Illiac IV.

calculators and mathematics textbooks for reference. In 1947, Harry Goett bought Ames' first electronic computer, a Reeves Electronic Analog Computer (REAC) and used it to drive simulators to study aircraft stability and control. The first digital computer, an IBM card program calculator, arrived in 1951. Ames' electrical staff lashed together three accounting machines from the IBM product line—a punch card reader, a printer, and an electronic calculator—and taught it to do mechanical reduction of wind tunnel data. To make better use of this machine, in 1952, DeFrance formed an electronic computing machines division, led by William Mersman. By 1955 Mersman's division had succeeded in connecting an Electrodata Datatron 205 computer directly to the 6 by 6 foot tunnel and the Unitary plan tunnels, making it one of the first computers to do real-time compilations of test results.

Harvard Lomax pioneered computational fluid dynamics (CFD) for use in solving complex aerodynamic problems.

Marcie Chartz (second from right) and Smith DeFrance (second from left) discuss the Ames installation of an IBM computer with IBM staff.

Now, tunnel operators could see almost immediately if their setup generated errors that required rerunning a test.

For seventeen years, Harv Lomax shared a carpool with Marcie Charz Smith, a woman computer who joined Mersman's division and who later became chief of the computer systems and research division. One morning, Lomax complained about having to redo a hand calculation because he used the wrong integral. Once at work, Smith wrote a one-line equation, pulled priority on the IBM calculator, and Lomax had his answer by eight o'clock that morning. Lomax became an instant convert, though other Ames theoreticians remained unconvinced that computers were here to stay. That changed in 1958 when Ames acquired an IBM 704 digital computer capable of running the Fortran programming language, with which they could calculate area rules that reduced drag on wing-body configurations. Calculations were a batch operation, done in octal dumps, meaning they did not know until after the punch cards finished running if there was a programming fault. So Lomax hooked up a cathode ray tube so he could watch the transactions in process and could stop the run if he saw a fault.

Ames opened its first dedicated, central computer facility (CCF) in 1961 adjacent to the circle ringing the headquarters building. At the heart of the CCF was a Honeywell 800 which replaced the Datatron and, until it was retired in 1977, collected data from all the wind tunnels for on-line data reduction.

The CCF also included an IBM 7094, used primarily for theoretical aerodynamics. Ames took its first step toward distributed computing in 1964 by adding an IBM 7040 to front-end the 7094 so that the time-consuming input-output efforts were not done directly on the 7094 computer processor. Ames acquired two smaller, short-lived mainframes—an IBM 360/50 in 1967 and an IBM 1800 in 1968. Mainframe computing took a giant leap forward in 1969, when Ames acquired an IBM duplex 360/67 as surplus from the Air Force Manned Orbiting Laboratory project in Sunnyvale. Now on one time-shared computer, Ames did scientific computing, administrative data processing, and real time wind tunnel data reduction. By adding remote job entry stations around the Center, Ames cut its teeth on distributed interactive computing.

Cray 1S

The Illiac IV originally had been built as a research tool in what was then called non-von Neumann computer architecture, and later called parallel processing. Burroughs Corporation built it, with funds from the Defense Advanced Research Projects Agency (DARPA), based on a design by Daniel Slotnick of the University of Illinois, for installation in the computer science department at the Urbana-Illinois campus. However, student unrest at campuses around the country, especially at the University of Illinois, made DARPA want to put the Illiac somewhere more secure. When Hans Mark heard through his old friend, Edward Teller, that the Illiac was in play, he asked Dean Chapman, new chief of the thermo and gas dynamics division, and Loren Bright, director of research support, to negotiate an agreement that got the Illiac

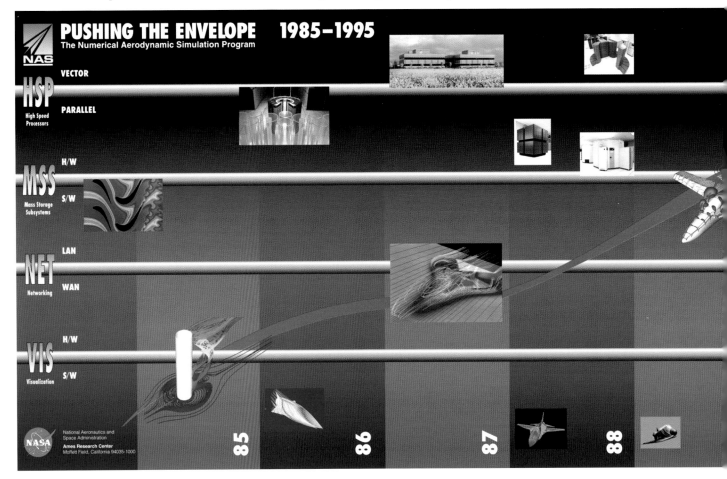

Cray X-MP

sited at Ames. Chapman and Bright promised that Ames could not only get the Illiac to work and prove the concept of parallel processing, but in the process would get a return on DARPA's $31 million investment by generating applications in computational fluid dynamics and in computational chemistry.

The Illiac IV arrived at Ames in April 1972. It was the world's first massively parallel computer, with 64 central processing units, and was the first major application of semiconductor rather than transistor memory. For three years, the Illiac was little used as researchers tried to program the machine knowing the results would likely be erroneous. In June 1975, Ames made a concerted effort to shake

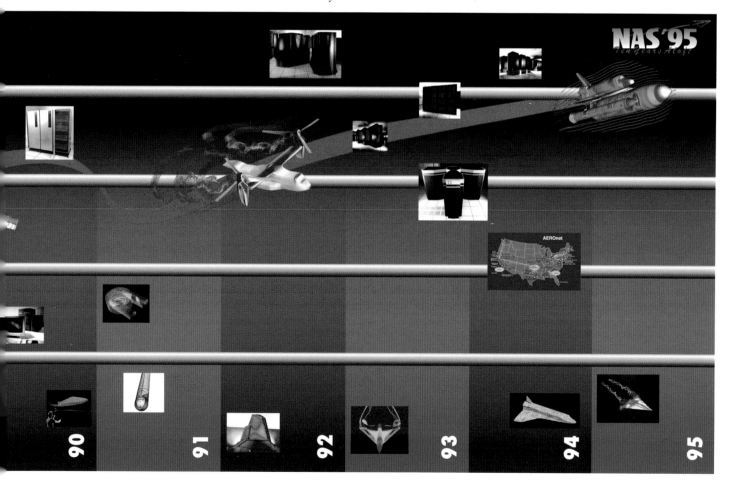

out the hardware—replace faulty printed circuit boards and connectors, repair logic design faults in signal propagation times, and improve power supply filtering to the disk controllers. Not until November 1975 was it declared operational, meaning the hardware worked as specified, but it remained very difficult to use. Designed for research in computer science, it lacked even the most primitive self-checking features. The programming language Burroughs wrote for it, called GLYPNIR, was general enough for computer science research but too bulky for efficient computational fluid dynamics. Most CFDers at Ames found it easier to continue writing Fortran codes and running them on existing serial computers. A few persisted, however. Robert Rogallo began looking at the architecture and the assembly language of the Illiac IV in 1971, even before it arrived. In 1973, he offered a code called CFD that looked like Fortran, and could be debugged on a Fortran computer, but that forced programmers to take full advantage of the parallel hardware by writing vector rather than scalar instructions.

Vector computing meant that programmers wrote algorithms that divided a problem into simultaneous discrete calculations, sent them out to the Illiac's 64 processors, then merged the results back into a single solution. Some problems in CFD were especially amenable to parallel processing. For example, airflow over a wing could be divided into cubic grids—containing air of specific temperatures and pressures—and the algorithms could compute how these temperatures and pressures change as the air moves into a new grid. Ames acquired a CDC 7600 computer in 1975, built by Seymour Cray of the Control Data Corporation (CDC) and also surplused from the U.S. Air Force. In translating Illiac-specific CFD language to run on the 7600, Alan Wray wrote

Cray-2

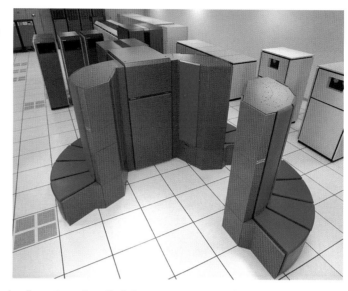
Cray Y-MP computer installed in the Numerical Aerospace Simulation facility, September 1988.

VECTORAL, a more general programming language used in some form in all subsequent supercomputers at Ames.

Ames had signalled its commitment to the development of parallel computing, and from then on the supercomputers arrived in a regular flow. Ames installed the Cray 1S in 1981, followed by the CDC Cyber 205 in 1984 (the largest ever constructed), the Cray X-MP/22 in 1984, and the Cray X-MP/48 in 1986. In addition, Ames was the launch customer for a variety of mini-supercomputers introduced in the early 1980s—like the Convex C-1, the Alliant FX/8, the Intel Hypercube, and the Thinking Machines Connection Machine.

All these computing tools attracted computing talent. In June 1983, James Arnold and Kenneth Stevens of Ames' astrophysics division formed the Research Institute for Advanced Computer Science (RIACS), allied it with the Universities Space Research Association, and recruited Peter Denning as its director. RIACS was designed as a bridge between Ames, the local universities and the computer industry. RIACS forged a match between the scientific problems of interest to NASA and the potential of new massively parallel computers, then created efficient new algorithms to solve kernel problems in CFD and computational chemistry. Ames researchers focused on theory, while visiting scholars at RIACS pioneered applications.

Cray C-90

By the mid-1980s, Ames was one of the world's leading centers in graphical supercomputing, massively parallel processing, and numerical aerodynamic simulation. To give these efforts a physical center, in March 1987 Ames opened the Numerical Aerospace Simulation facility, called the NAS. At the heart of the NAS was one of the world's greatest central processors, the Cray-2 supercomputer. The Cray-2 had an enormous 256 million word internal memory—sixteen times larger than any previous supercomputer—because Ames CFDers had visited Seymour Cray to impress upon him the need for massive memory that was quickly addressable. It was the first Cray to run the Unix operating system, the emerging open standard in scientific and university computing, which brought new blood into the field of CFD. It had cost $30 million, computed a quarter of a billion calculations per second, and had to be cooled by liquid nitrogen rushing through clear plastic tubes. Ames had acquired the Cray-2 in September 1985, and had already written the technical specification for the computer that would supersede it. The Cray Y-MP arrived in August 1988, sporting eight central processors, 32 megawords of central memory and a $36.5 million price. The Y-MP performed so much better because its bipolar gates allowed faster access to memory than the Cray-2's metal oxide semiconductor memory. The NAS plan was to always have in operation the two fastest supercomputers in the world. By May 1993 the NAS added the Cray Y-MP C90, then the world's fastest, and six times faster than the Y-MP.

The NAS building itself was sophisticated. As a home for the Cray, it was kept cool and clean by an air system thirty times more powerful than the systems serving any normal office building of 90,000 square feet. NASA expected to fund ongoing operations at the NAS with an annual appropriation of about $100 million, so the NAS also housed one of the world's great computer staffs and a range of input and

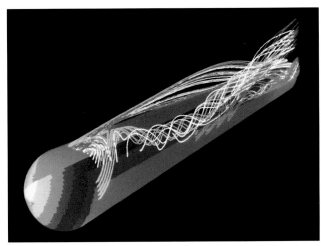

High alpha flow about a hemisphere cylinder.

NAS Facility

Airflow around an AV-8B Harrier.

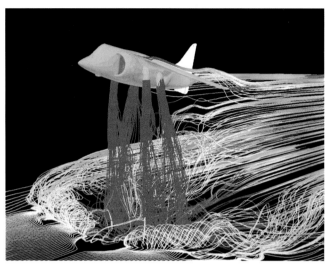

Flow structures between the wing and body of a McDonnell Douglas F-18 Hornet.

output devices. Support processors had friendly names, like Amelia, Prandtl, and Wilbur—the smaller processors named for aviators, the larger ones for mathematicians. The NAS acquired the earliest laser printers, and pioneered the development of graphical display technologies. F. Ron Bailey, the NAS project manager, directed the NAS to provide supercomputing tools for aerospace research as well as to forge the development of computing itself.

Though the NAS was a physical center for computing at Ames, its tentacles reached into much larger communities. First, around Ames, NAS staff worked directly with the wind tunnel and flight researchers to make CFD an important adjunct to their work. Virtually every other research community at Ames—those working in the life, planetary, astronomical, and materials sciences—found the staff of Ames' computational chemistry branch ready to find new ways to apply supercomputing to research questions. Plus, the NAS was wired into the larger world of science. ARPA had decided that its Illiac should be accessible via the Arpanet—a network of data cables that linked universities and national laboratories. Hans Mark agreed, based on his experience in using supercomputers in the nuclear laboratories following the discontinuance of above-ground tests. Editors, compilers, and other support software for the Illiac ran only on IBM, DEC, or Burroughs computers. Programmers submitted their code while remotely logged into the IBM 360, usually between the hours of midnight and eight o'clock in the morning, and

Diverse Challenges Explored with Unified Spirit: 1969 – 1989

results were returned back over the Arpanet. This made the scientific community more aware of bandwidth and reliability limitations of the network, and Ames continued to lay cables leading to the Arpanet ring around the Bay Area.

Networking grew stronger as computing pervaded every area of research at Ames. Budget pressures in the mid-1970s forced Ames to do more with less. Jim Hart, on the technical staff of the computation division, encouraged Ames research groups to acquire smaller, interactive (non-batch) computers, with graphics capabilities, and to link them together. Beginning in 1978, Ames acquired several VAX computers from the Digital Equipment Corporation (DEC) and soon Ames had the largest DECnet in the world—outside of the DEC corporation itself—and a reputation for aggressive development of distributed computing. In November 1982, Ames computer scientists Eugene Miya, Creon Levit and Thomas Lasinski circulated an electronic mail message asking "What is a workstation;" specifically, how a workstation should divide the many tasks of scientific computing with the network and the mainframe. They compiled the comments into the specifications for the first graphic design workstations built by local firms with close ties to Ames—Sun Microsystems and Silicon Graphics, Inc.

By the mid-1980s, however, the dedicated computer-to-computer wiring of the DECnet was being superseded by the packet-switching TCP/IP data transfer protocol that drove the explosion of the Internet. So Ames made a commitment to the technology that allowed closer collaboration with universities and industry: TCP/IP, servers running the UNIX operating system as refined by Silicon Valley firms

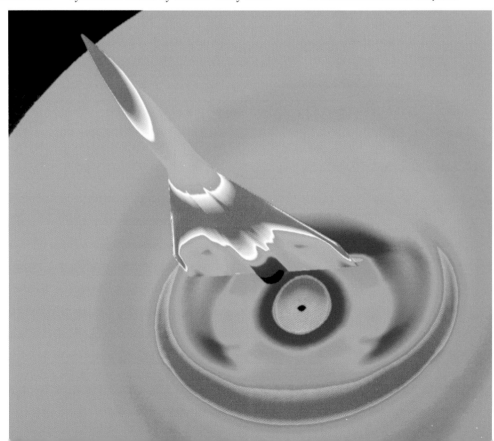

Predicted sonic boom footprint.

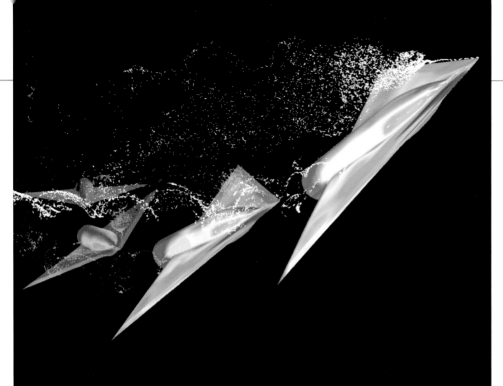

Delta wing roll motion.

like Sun Microsystems, and object-oriented client computers like the Apple Macintosh. In 1989, NASA headquarters asked the Ames central computing facility to form a NASA Science Internet project office (NSI) which would merge NASA's DECnet-based network into a secure TCP/IP network. By the time the Cray Y-MP was operational in 1989, more than 900 scientists from 100 locations around the United States were wired into the NAS over the Internet.

Computational Fluid Dynamics

The technology of computational fluid dynamics (CFD) is transferred via computer codes—generic programs into which aerospace designers enter a proposed design in order to model how air flows around it. The increasing sophistication of these codes—over the two decades Ames committed itself to CFD—reflected not only the application of greater computing power, but also a concomitant flourishing in aerodynamic theory around the Navier–Stokes equations.

The Navier–Stokes equations were introduced in 1846, as a theoretical statement coupling various algebraic equations based on the rules of conservation of mass, momentum and energy. The Navier–Stokes equations are so complex that until the advent of CFD aerodynamic theorists avoided the full set of equations. Aerodynamicists won acclaim, instead, by reducing a flow calculation to its essence and then applying the appropriate partial differential equations—either elliptical, hyperbolic, or parabolic. The only flows they could simulate were for slender aircraft, at small angles of attack, outside the transonic regime, flying in perfect gas with no viscosity and with no flow separation. Thus, even though the advent of Fortran-based computers in the 1960s made it possible to run these so-called inviscid linearized equations in three dimensions, the simplified configurations on which their calculations were based bore little resemblance to actual aircraft. Nevertheless, Harvard Lomax continued to refine his calculations of supersonic flows over blunt objects, and Robert MacCormack of the vehicle environment division continued to refine his calculations of viscous flows.

Shuttle in launch configuration showing surface pressure comparisons.

In the early 1970s, CFD took a major leap forward with codes that allowed the velocity, density, and pressure of air flowing over a realistic aircraft design to be calculated, ignoring only viscosity or flow separations. Ames CFDers wrote codes that generated results near Mach 1 and other speeds where tunnel data were unreliable—codes to model wing-body interactions in transonic flow, the blast wave over a hypersonic missile, blunt bodies, and supersonic aircraft configurations. The first experiment run on the Illiac IV was a model of how a sonic boom changes as it approaches ground air. Thomas Pulliam wrote the ARC3D code, which superseded Harvard Lomax's ARC2D code. For the first time, the Illiac allowed three-dimensional portrayals of airflows.

By the late 1970s, with the Illiac IV in more routine operation, CFDers were modeling incompressible flows—flows in which the atmosphere expands or grows denser, adding kinetic energy to the flow and requiring equations that couple velocity and pressure with temperature. This was the first step toward models of supersonic and hypersonic shock waves, as well as models of turbulent boundary layers. By the early 1980s, CFDers had essentially developed a complete set of Navier–Stokes solvers. They had computed time-dependent flows, which depicted how flows changed over time, rather than time-averaged flows, which showed their general tendencies. Furthermore, they had improved their models of turbulence, from simple eddy viscosity models to finite difference models of turbulence in separated flows. Some, like Helen Yee, worked on using nonlinear chaos theory to study turbulence numerically. Ames and Stanford University, in February 1987, formed a joint venture called the Center for Turbulence Research to specifically develop turbulence models to inject into the Navier–Stokes equations. Once these individual calculations were proved theoretically, Ames CFDers coupled them together to push the Navier–Stokes equations to the limits of their approximation. They also packaged them into routine codes with real industrial significance.

At first, CFDers used tunnel data to validate their computed results. Then, CFDers wrote code that complemented tunnel tests by modeling flows that were impractical to test in a tunnel. Eventually, CFD replaced tunnel tests by generating results that were cheaper and more accurate than data obtained in a tunnel. As airframe companies made more complex aircraft, the number of tunnel and flight tests required in the design of any new aircraft grew at an exponential rate in the 1960s and 1970s. Charles "Bill" Harper, who led Ames' full-scale and systems research division, made this argument in a major 1968 address. During F-111 design definition, in the mid-1960s, Ames did 30,000 hours of tunnel tests at a cost of $30 million. For the Space Shuttle, Ames aerodynamicists planned even more tunnel time. CFD codes, they expected, could eventually eliminate half of this testing in the early design stage.

The first major research program at the NAS validated the design parameters for the National Aerospace Plane, a Reagan administration effort to build an aircraft that could take off from a runway and reach low-Earth orbit. Using the Cray-2, Ames researchers evaluated airframe designs proposed by the three contractors, calculated thermal protection requirements, and suggested ways of integrating the unique scramjet engine into the shock waves around the airframe. Ames' computational chemistry branch helped by calculating the energies released by air-hydrogen combustion and by evaluating the promise of ceramic composite heat shields. Of course, others at Ames then validated all these computational results with tests in the wind tunnels or in the arc jet complex.

Unsteady multistage turbomachinery flows.

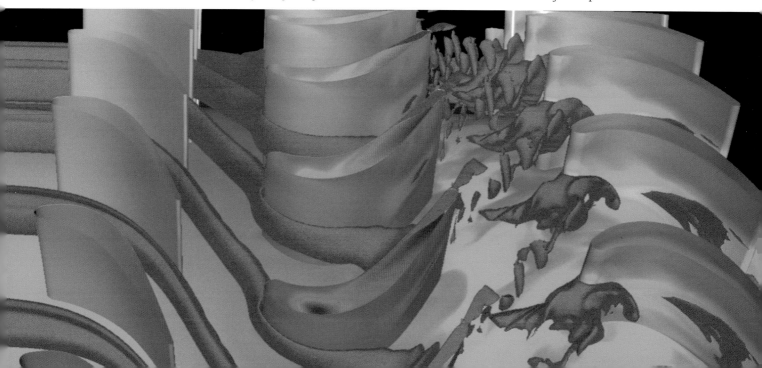

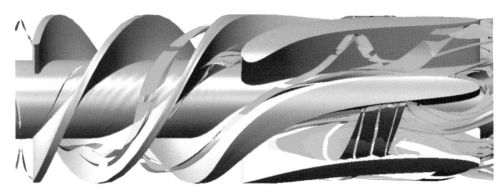

Flows inside the propeller of a left ventricular assist device.

Thus, in less than two decades, Ames had brought the field of CFD to maturity. Ames people helped design the supercomputers, visualization equipment, and internetworking that linked them. Ames people rebuilt aerodynamic theory around the complete Navier–Stokes equations, wrote the codes for general proximations of airflow, rendered these codes routine design tools, then pioneered codes for more complex problems. Ames CFDers authored code for virtually every flow problem: external as well as internal flows in the subsonic, transonic and hypersonic regimes. And they coupled these codes to encompass more parts and, eventually, to model entire aircraft and spacecraft. Ames CFDers then worked up theories of numerical optimization, so that designers could specify the performance of a new design and the code would define the best configuration for it. Wing designs, especially, could be optimized computationally so that wind tunnel tests were needed only to verify this performance.

Ames CFDers wrote codes used in the design of virtually every aircraft in the western world. The Cray version of ARC3D was reportedly used to hone the first Airbus, the A300. Ames developed the general aviation synthesis program (GASP) to do quick configuration studies of general purpose aircraft. Industrial users included Beech Aircraft, Avco–Lycoming, and Williams International. The code was used to analyze configurations of subsonic transport aircraft with turbo-props, turbofans, prop-fans, or internal combustion engines. It predicted flight performance, weight, noise, and costs, and allowed easy trade-off studies. Ames CFD work helped Orbital Sciences, a start-up company trying to develop the first new American launch vehicle in two decades. Under NASA's program for small expendable launch vehicles, Ames CFDers adapted code to hone the design of Orbital's air-launched Pegasus rocket and arranged for flight tests with the Pegasus hanging under the Ames–Dryden B-52 aircraft in November 1989. Boeing and McDonnell Douglas closely followed the state of the art in CFD to refine their commercial transports, but by far the biggest users of CFD were entrepreneurial firms or the airframe firms designing entirely new fighter aircraft.

For designers of supersonic inlets, Leroy L. Presley of Ames devised the first three-dimensional internal flow code. For rotorcraft designers, including those at Ames working on VTOL aircraft, Ames CFDers devised various computer codes to model the complex aerodynamics of helicopters. CAMRAD was a comprehensive code capable of analyzing various rotor configurations—tandem, counterrotating, and tilt rotor—used to predict blade loads, aeroelastic stability and general performance. ROT22 was a code for rotor field flows, applicable from hover to forward flight, and was three-dimensional, transonic, and quasi-steady.

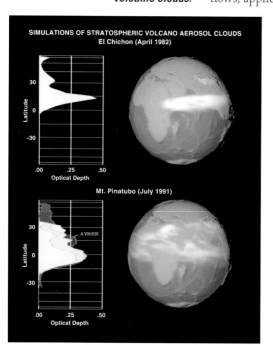

Simulation of stratospheric volcanic clouds.

In 1988, Ames researcher Man Mohan Rai published a code to model the complex pressures, temperatures, and velocities within a jet turbine engine. Engine parts move constantly relative to one another, clearances are very tight, and pressure changes produced by entering air creates unsteady states. Controlled experiments of engine concepts with physical prototypes were very expensive. Rai's model not only solved unsteady three-dimensional Navier–Stokes equations, but did so for complex geometries. It initially required 22 trillion computations, performed on the Cray X-MP at the NAS, before others at Ames set to work simplifying the code to make it a practical tool for industrial design. A highly accurate method for transferring calculated results between multiple grids was the key to Rai's model, and this method later found extensive applications to multiple rotor-stator aircraft.

Some NAS programmers applied their codes to the solution of peculiar problems which then shed light on more general solutions. To depict flows within the space shuttle engines, Ames CFDers Dochan Kwak, Stuart Rogers and Cetin Kiris created a program called INS3D (an incompressible Navier–Stokes solver in general three-dimensional coordinates). Because it was useful in modelling low-speed, friction-dominated flows, in 1993 the group also applied the code to model airflow over transport aircraft at takeoff and to improve a mechanical heart developed at Pennsylvania State University.

Not all of Ames supercomputing focused on modeling airflows. In fact, only twenty percent of the computing time on the Illiac IV was spent on aerodynamic flows. A wide flung group of users, overseen by Melvin Pirtle of the Institute for Advanced Computing, spent the rest of the time on modeling climates, seismic

Bill Ballhaus was a leading proponent of Ames' Numerical Aerospace Simulation facility.

plate slippage, radiation transport for fission reactors, and the thermal evolution of galaxies. When the NAS became available, Ames people wrote codes using maximum-likelihood estimation theory to extract aerodynamic stability derivatives from flight data. Airframe designers worldwide used this code to acquire aircraft parameters from flight data, and thus validate aerodynamic models, update simulators, design control systems and develop flying qualities criteria. Ames people wrote the hidden-line algorithms underlying most computer-aided design. This code depicted large, complex, engineering renderings faster than ever, and could be applied to aircraft design, architecture and systems design. It became the best-selling software in NASA history. But by far the biggest nonaerodynamic use of the Ames supercomputers was for computational chemistry.

Computational Chemistry

Aerothermodynamics and heat shield research brought computational chemistry to Ames. James Arnold had spent several years analyzing the chemical properties of shock-heated air and other planetary gases, and how these atmospheres interacted with ablating materials on heat shields. In 1969 Hans Mark challenged him: "Why don't you compute gas properties, rather than relying on measurement?"[7] Ames had done superb work building shock tunnels and simulators for atmospheric

Paul Kutler (right) guided much of Ames' work in computational fluid dynamics.

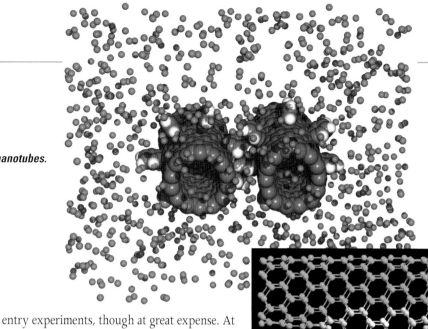

Nanogears and nanotubes.

Henry Lum directed Ames' early work in intelligent systems.

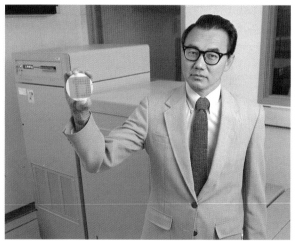

entry experiments, though at great expense. At spectroscopy meetings, Arnold heard of work at Argonne National Laboratory that showed the potential for reliable computations of the gas properties of small molecules. With his colleague Ellis Whiting, Arnold saw ways to apply Ames' emergent infrastructure in supercomputing to solve problems in atmospheric entry physics. They were supported by Mark and by Dean Chapman, who had pioneered the theory of aerothermodynamics and later, as director of astrophysics, helped lead Ames into computational solutions. Ames' computational chemistry branch developed, under Arnold's leadership, into a unique resource in NASA.

Academic chemists had computed results that were accurate only for single atoms. Fairly quickly, computational chemists at Ames—including Stephanie Langhoff, Charles Bauschlicher, and Richard Jaffe—developed tools to predict rates of gas-solid chemical reactions involving thirty atoms, predicted forces in molecules and atomic clusters as large as 65 atoms, and simulated material properties involving up to 10,000 interacting atoms. Applying this work to problems of interest to NASA, they designed polymers that were resistant to degradation by atomic oxygen, and improved noncatalytic thermal protection systems. Computational chemists explored several species of ablative materials for the heat shield of the Galileo probe—which had to be well matched to the atmosphere of Jupiter—and derived the radiative cross sections and absorption coefficients of these species

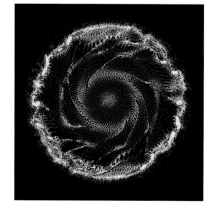

Interacting ring galaxies.

Diverse Challenges Explored with Unified Spirit: 1969 – 1989

Carol Stoker with the NASA Mars underwater rover and telepresence test bed, January 1992.

to determine which data were required to design the heat shield.

With these tools in place, David Cooper then led the Ames computational chemistry branch to apply its research to other problems. To develop better aircraft fuels, Ames explored the chemistry of transition metals used in catalysts. To develop better gas properties for aircraft engine flows, Ames computed bond energies and gas transport properties more precisely than ever done experimentally. To develop smaller robotic vehicles, better computer memory devices and other nanotechnologies, Ames calculated how to make materials bond at the molecular level. To understand the chemical evolution of the solar system, Ames calculated the composition of unidentified spectra observed from space telescopes. Within a decade, Ames had nurtured computational chemistry into a discipline of major importance to American industry and NASA.

Most important, virtually the entire first generation of CFDers and computational chemists had circulated through Ames in order to use the best machines, to try out forthcoming codes, and to train with the best in the field. And as Ames computational experts saw their fields mature, they reinvented themselves as pioneers in new areas of information technology like artificial intelligence, virtual reality, real-time computing, and distributed networking.

Intelligent Systems and Telepresence

In the early years of artificial intelligence (AI), symbols rather than numbers were used to represent information, and heuristic rules structured this information rather than the yes/no algorithms used in numerical computation. In 1980 Henry Lum acquired a computer that ran the LISP (for list processing) computing language,

The Ames TROV (for telepresence remotely operated vehicle) during underwater trials before an Antarctic mission.

and used it to develop the symbolic language of artificial intelligence. Increasingly, Ames researchers focused specifically on communications protocols for integrating various artificial intelligence agents, as needed to guide complex spacecraft or manage complex and changing projects. The goal was to construct rational agents that can acquire and represent abstract and physical knowledge, and reason with it to achieve real world goals.

Ames formed an information sciences division in June 1987 to spearhead the application of artificial intelligence to space missions. NASA had plans for an autonomous Mars rover, and Ames hoped to provide the technology for many such intelligent agents. The enormity of NASA's just-announced Space Station, for example, required onboard automation for many of the housekeeping functions that would otherwise need to be done by astronauts. Ames' artificial intelligence branch looked at the scheduling of shuttle orbiter ground processing and developed software that, beginning in 1993, saved NASA $4 million a year in shuttle maintenance. "Shuttle refurbishing is a difficult problem because you can only predict half of the work in advance," noted Monte Zweben, who led a team of contractors at Ames and the Johnson Space Center, shared in the largest Space Act award ever granted by NASA, then left to start up a company to program scheduling software for industry.[8] Peter Friedland led a group working with Johnson to automate Shuttle mission control and reduce human-intensive tasks by forty percent. Silvano Colombano worked with MIT researchers to develop the

Marsokhod rover in the Ames sandbox during evaluations of an x-ray defractometer.

199

Field test of goggles designed for telepresence on Mars.

astronaut science advisor, a laptop computer running artificial intelligence software that helped astronauts optimize spaceborne experiments as they unfolded. Astronauts referred to it as the "P.I. in a Box"—like having the principal investigator on board. While the Ames information sciences division looked for ways to contribute to larger NASA missions, for missions not yet conceived they continued to refine the general principles of artificial intelligence.

Artificial intelligence is a key component in enabling humans and robots to work together as an integrated team of rational agents, when coupled with the technology of virtual reality and telepresence. In 1984, when Michael McGreevy, a researcher in spatial information transfer, learned that a head-mounted display developed for the Air Force would cost NASA a million dollars, he pulled together a team to build its own. The result was VIVED (for virtual visual environment display), the first low-cost head-tracked and head-mounted display, with stereo sound and a very wide field of view. McGreevy soon built the first virtual environment workstation by integrating a number of components, including the VIVED helmet, a magnetic head and hand tracker, a custom-built image conversion system, an Evans & Sutherland vector graphics display

Marsokhod Russian rover was a hardy platform for testing telepresence technology. In 1995 it simulated Martian terrain by exploring the Kilauea volcano in Hawaii.

system, a DEC PDP-11/40 computer, and software he wrote that generated and displayed three-dimensional, interactive, stereoscopic scenes of commercial air traffic in flight. It was the first major advance in wearable personal simulators since the laboratory systems built by Ivan Sutherland in the 1960s. By 1987 NASA had boosted the budget for this work thirtyfold.

A whole industry was built around virtual environments, with many of the major innovations inspired or filtered through Ames. Start-up VPL Research of Redwood City commercialized the VIVED design and supplied low-cost virtual reality systems around the

Image taken from the Ames C-130 of lava flows from the Kilauea Volcano, Hawaii.

Vic Vykukal testing the AX-5 spacesuit in the Ames neutral buoyancy tank, in August 1987, to determine the best technologies for spacesuits to be used aboard the Space Station.

world. Scott Fisher, who joined Ames' virtual reality team in 1985, worked with VPL to develop a data glove for computer input. Though the first systems at Ames used Evans & Sutherland vector graphics processors, Ames later used some of the first more powerful and affordable raster graphics systems. Jim Clark credits the many graphics projects at Ames with helping his start-up company, Silicon Graphics, Inc. of Mountain View, California, build image-specific tools and chips. Since the late 1980s, Ames and SGI have worked closely to advance the tools of image generation and virtual reality. Also, Ames work in virtual reality was possible only with new tools for real-time computing. Working with Sterling Software, an Ames support contractor, Ames people developed the mixture of peripherals and interfaces for data acquisition, telemetry, controls, computer animation, and video image processing to compute and portray data points as they were collected.

Cedi Snowden analyzes the AX-5 spacesuit glove.

Virtual reality put Ames at the forefront of human-centered computing. With human-centered

Diverse Challenges Explored with Unified Spirit: 1969 – 1989

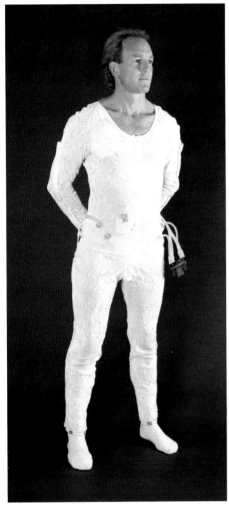

Liquid cooling garment, developed at Ames as part of its spacesuit research, worn by Phil Culbertson.

computing, people would not consciously interact with the computer itself, but rather interact directly and naturally with real, remote, computer-augmented or computer-generated environments of any kind. NASA saw the value it might have on the space station, by allowing astronauts to control robotic devices around the station. Ames used images generated by CFD to build a virtual wind tunnel—wherein the wearer could walk around a digitized aircraft and see the brightly colored lines depicting airflows. Elizabeth Wenzel of Ames' spatial auditory displays laboratory led a university and industry team developing "virtual acoustics" using headphones to present sounds in three-dimensions. Stephen Ellis and Mike Sims developed other key components of virtual reality. Ames saw other uses for it—in virtual planetary exploration. As NASA's planetary probes were digitizing the planets—like Magellan's mapping of the surface of Venus—

Virtual reality gloves and headgear, 1989.

Ames used those data to generate images projected through the personal simulator. It gave anyone—geologists, astronauts, journalists or schoolchildren—the feeling of

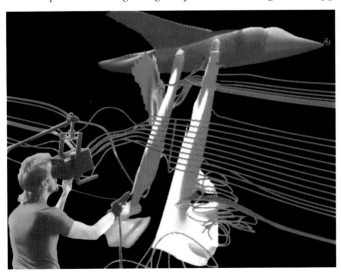

Steve Bryson, outfitted with virtual reality gloves and headset, displays the Ames virtual wind tunnel.

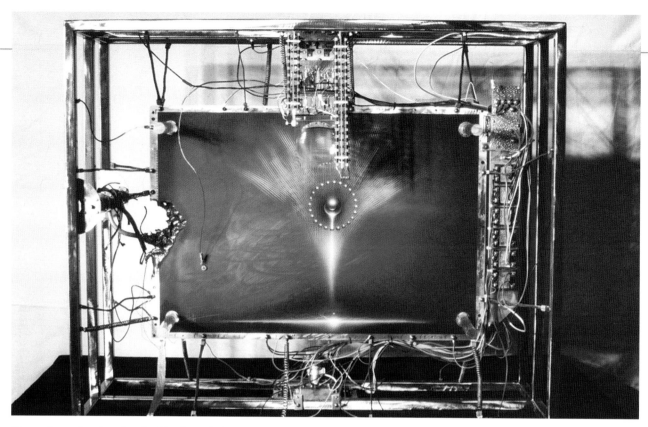

Three-dimensional art, inspired by the artist's experience at Ames. Andreas Nottebohm tested Ames' virtual reality headset as it showed a computer-generated scene or a real scene relayed by video cameras. This technology is meant to provide telepresence and telerobotics for exploration of other worlds.

being there. They used the panoramic views returned from the Viking landers to plan the digitization technology for the Mars Pathfinder, then tested this technology on remotely operated rovers. Prototype rovers imaged the hostile terrain around Death Valley, Antarctica, the volcanoes of Alaska and Hawaii, and underwater in the Monterey Bay. The Marsokhod Rover, lent to Ames in 1993, was a superb platform on which to test this capability called telepresence.

Work in human-centered computing at Ames took a major leap forward in 1989 with the opening of the human performance research laboratory (HPRL). David Nagel had championed the laboratory to house Ames' aerospace human factors research division. After all, Ames' long tradition of work in flight simulators and fly-by-wire technology was a form of telepresence. In addition to supporting Ames' longstanding work in aviation flight training, cockpit resources, and pilot and controller performance, the HPRL brought together researchers working to solve the problems of extended human presence in space, like Vic Vykukal's work in spacesuit design. There, Ames continued its work on making spacecraft more habitable for long-term residents, by investigating microgravity restraints, visual orientations, and changes to circadian rhythms. "We consider it our responsibility to not only promote the productivity of people housed in space," noted Ames environmental psychologist Yvonne Clearwater, "but to assure that once there, they will thrive, not merely survive."[9]

Built adjacent to the human factors laboratory was the automation sciences research facility (ASRF) so that experts in human factors and artificial intelligence could collaborate. The ASRF opened in January 1992, four months ahead of schedule and $500,000 under its $10 million budget. The ASRF provided office space for the growing numbers of artificial intelligence and robotics experts at Ames, led by information sciences division chief Henry Lum. It also provided eleven superb laboratories. In the high bay, Ames built a simulated lunar terrain and used it to test intelligent systems for a rover that would explore planetary surfaces.

CONTINUING DIRECTION: WILLIAM F. BALLHAUS, JR.

The inculcation of supercomputing into everything Ames did accelerated when Bill Ballhaus, a leader in CFD, became Ames' next director. By 1984, Sy Syvertson had directed Ames for six years, and the Center had flourished under his guidance. But the death of some close friends on the Ames staff, a series of heart problems, and the tragedy and inquiry following the accident in the 80 by 120 foot wind tunnel, all caused him to think it was time for younger leadership. He encouraged headquarters to look at Bill Ballhaus, who had already distinguished himself as a leader.

Ballhaus received his B.S., M.S. and Ph.D. degrees from the University of California at Berkeley in mechanical engineering. His father was a senior vice president for Northrop Aerodynamics and Missiles in Los Angeles, and introduced him to the emergent importance of computing in aerospace. Ballhaus served in the U.S. Army Reserve from 1968 to 1976, earning the rank of captain. He arrived at Ames in 1971 as a civil service engineer with the U.S. Army Air Mobility Research and Development Laboratory. When Ames decided to form an applied computational

Space Station simulator, during research on how best to manage a shuttle docking.

William F. Ballhaus, Jr., Director of Ames from 1984 to 1989.

aerodynamics branch, the Army staff was delighted to let Ballhaus become a NASA employee as branch chief. It proved that a close working relationship had developed between the Army and Ames. After a year, Ballhaus became Ames' director of astronautics in 1980. CFD underwent explosive growth in the 1970s, and Ballhaus honed his leadership skills through almost constant recruitment. Along with his younger colleagues in the field—Paul Kutler and Ron Bailey—Ballhaus kept abreast of work done in industry and academia, learned to quickly size up whether a researcher wanted time to do basic research or the excitement of engineering application, and teamed them with the best colleagues.

Ballhaus became the director of Ames in January 1984, and helped bring on line several facilities that were key to its research future, like the Numerical Aerospace Simulation facility and the NFAC. Ballhaus initiated Ames' first comprehensive strategic planning exercise, published in March 1988, which suggested that information technology could inject new life into every research area at Ames. And Ballhaus was skilled in reading headquarters, helping Ames people sell their research efforts by describing their ultimate contributions to the International Space Station. Funding for Station-oriented projects was good, and the Ames budget grew quickly in the late 1980s.

Four years into his directorship, in February 1988, Ballhaus was called to Washington to serve one year as acting associate administrator for NASA's Office of Aeronautics and Space Technology. This made him responsible for the institutional management of the Ames, Langley and Lewis research centers. Once NASA named a permanent associate administrator of OAST, Ballhaus returned as Ames director, but stayed less than six months; on 15 July 1989 he officially resigned. He insisted that the press release about his resignation cite "inadequate compensation for senior federal executives and vague new post-government regulations as factors in his decision."[10] This referred to a 1989 ethics law that barred federal contractors from

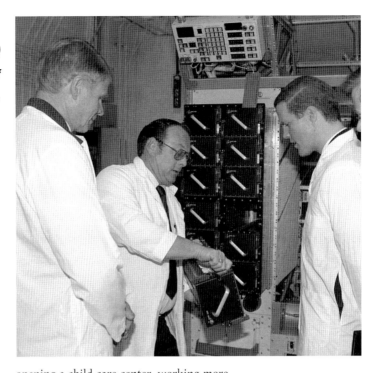

Dale Compton (left) and Bob Hogan (center) giving Congressman Tom Campbell a tour of Ames' life sciences laboratories.

hiring federal employees who had supervised their competitors' projects. Ballhaus was one of several NASA officials to leave the agency in the week before the new law took effect, prompting the newly appointed NASA Administrator Richard Truly to call a press conference to decry the law as "a crying shame."[11]

Throughout his tenure as Ames' director, Ballhaus amplified a concern expressed by all previous directors—that Ames needed the freedom to hire the best people. Back in October 1961, when Vice President Lyndon Johnson asked Smith DeFrance what he could do to help Ames, Defrance asked for freedom from civil service hiring ceilings. The ceilings remained an issue, and Ames was never so constrained by funds or resources as it was by civil servants to manage them. By the 1980s, Ames still suffered under the ceilings, but now lacked the freedom to pay potential hires competitive wages. Ballhaus fought to secure special salary rates applicable to half of the Ames workforce, limited approval to match industry salary offers, hiring authority for most of the occupations at Ames, and approval to test out a more flexible compensation and promotion plan. He led his staff in improving the quality of life around Ames—opening a child care center, working more closely with the National Federation of Federal Employees, getting everyone involved in a regular strategic planning process, and encouraging diversity so that Ames was awarded the NASA trophy for equal employment opportunity in both 1984 and 1989. Statutes limited what he could do with executive pay, however, and when Congress defeated the Reagan administration proposal for a pay raise many in Ames' senior executive service left prematurely. "I would have preferred a more graceful exit," Ballhaus wrote to announce his departure. "The Center's success in the future will depend upon our ability to continue to recruit and retain the high-quality people that Ames is noted for. In leaving, it is the close association with the outstanding people who make up this Center that I will miss most."[12] From there Ballhaus joined the Martin Marietta

Astronautics Group in Denver as vice president of research and development, then rose steadily up the ranks of Lockheed Martin Corporation.

Dale L. Compton

Dale Compton, who had served as acting director when Ballhaus moved to Washington, replaced him as Ames' director. Compton, too, was a product of Ames. He came to the Center fresh out of Stanford University with a master's degree in 1958, one of the first students taught by former Ames aerodynamicist Walter Vincenti. He then returned to receive his Ph.D. in 1969. Compton worked as an aeronautical engineer and had a penchant for participating on project teams—as an aerothermodynamicist for ballistic missiles and NASA's Mercury, Gemini and Apollo human space programs, and as manager of the infrared astronomical satellite program (IRAS). He entered management ranks in 1972 as deputy director of astronautics, became chief of the space sciences division, became director of engineering and computer systems, was named Ballhaus' deputy in 1985, and was officially named director on 20 December 1989 at ceremonies marking Ames' fiftieth anniversary.

Victor L. Peterson joined Compton as deputy director in 1990. Peterson, too, was a product of Ames. He joined Ames in 1956 upon graduating from Oregon State University, and distinguished himself through research in aerodynamics, high-temperature gas physics, and flight mechanics. He was internationally known as an advocate of large scale scientific computing in all scientific disciplines, but especially in computational fluid dynamics.

Compton, like Ballhaus and Syvertson before him, understood how Ames nourished innovation and personal reinvention. Each had grown his own career at Ames,

Dale L. Compton, Director of Ames Research Center from 1989 to 1994.

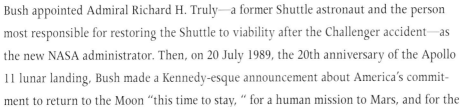

David Wettergren and Natalie Cabrol control the Nomad, a platform built at Carnegie-Mellon University, during the Nomad's trek through the Atacama Desert.

and each knew how to let those under his direction blossom. And NASA headquarters provided new opportunities and resources for myriad Ames researchers to flourish as the Bush administration looked to space adventures—following the end of the Cold War in 1989—to once again display America's technological prowess.

In April 1989, early in his term as president, George Bush appointed Admiral Richard H. Truly—a former Shuttle astronaut and the person most responsible for restoring the Shuttle to viability after the Challenger accident—as the new NASA administrator. Then, on 20 July 1989, the 20th anniversary of the Apollo 11 lunar landing, Bush made a Kennedy-esque announcement about America's commitment to return to the Moon "this time to stay, " for a human mission to Mars, and for the expanded internationalization of the Space Station Freedom. These long-term, complex space projects made good use of the basic research done at Ames in microgravity, robotics, and planetary science, and Ames' budget grew apace modestly into the early 1990s.

Yet Compton was seen by some around Ames as too conservative in his vision—a "tunnel hugger"—one who thought Ames' position within NASA depended on the immovability of the superb wind tunnel infrastructure around Ames. Compton had seen the more project-oriented NASA Centers go through booms and busts

On 20 December 1989, Ames buried a time capsule and unveiled a sculpture at the spot where, fifty years earlier, Russell Robinson had turned the first spade of dirt for the Ames construction shack: Robinson (left), Compton (center), and Syvertson (right).

as Congress approved and disapproved major projects and thought Ames—fundamentally a basic research organization—would be especially disrupted by such cycles. He had doubts about what sort of institutional follow-on would come from any of the projects emanating from Ames' space scientists, and he understood that if the Jet Propulsion Laboratory needed work that NASA headquarters would send space projects there to be managed. He had fought hard for SIRTF (the space infrared telescope facility), the Mars Observer, and the Magellen Venus all to be managed at Ames, but all were lost to JPL. As deputy director, Compton had nurtured the airborne telescope SOFIA only to see, as director, it cut at the last minute before submission of the final NASA budget. Moreover, the various wind tunnel and simulator restoration projects added $300 million to Ames' budget in the late 1980s, so Compton made sure these efforts were managed well.

Beginning in the late 1980s and continuing through the mid-1990s, NASA headquarters put Ames through a series of roles and mission exercises. The goal, ultimately, was to make all NASA Center directors more agile in being able to modify their Centers' expertise to accommodate changing national needs. While the strategic plans emerging from these exercises always reiterated Ames' interest in aeronautical research, the plans always seemed a bit empty. A great many people at Ames, especially those in life sciences and information technology, began to wonder how they fit into that picture of Ames. Into the 1990s, Ames began to directly address the relationship between its future and its past.

Human centered computing is on the forefront of Information Technology at Ames.

1990   1999

A Center Reborn

Chapter 4:
Ames in the 1990s

Ames underwent more profound change in the mid-1990s than in any period since the end of the Apollo era. With the demise of the Soviet threat and shrinkage in federal research spending, Ames people had to face the reality that their Center might be shut down. Like NASA as a whole, Ames was swept up in changes imposed by headquarters: downsizing, quality management, reengineering, program shifting and outsourcing.

However, Ames people took this dark period as an opportunity for self-discovery—of asking what was unique about Ames' historic strengths in science and engineering. They focused on expansive new missions in astrobiology and intelligent systems, and cleared away inherited structures to get at the essence of their work. By the end of the decade, as NASA as a whole reconfigured itself to shape America's aerospace future, the Ames approach—its cultural climate, managerial empowerment, collaborative spirit and fundamental scientific curiosity—increasingly stood as the model for what NASA wanted to become.

Calothrix cyanobacteria isolated from Midway Geyser Basin in Yellowstone National Park. Analogous thermal spring features have been identified on Mars and are of interest as potential landing sites to search for ancient life.

THE GOLDIN AGE

Three years into the Bush administration, Congress insisted more firmly that all federal laboratories, especially those in the departments of energy and defense, rethink their roles for the political realities of the post-Cold War era. Compared with the rest of NASA, Ames had lost little as Congress started cutting defense funds. Ames had already made plans to mothball all nonessential tunnels and simulators. Half of Ames' remaining tunnel time went to test military aircraft, though civil projects stood in line to buy any freed-up time. What military work that remained at Ames went toward technologies—like helicopters and navigation systems—needed to fight the now-expected strategic scenario of many battles on many fronts. In fact, the decades of quiet collaboration between Ames and the Soviets in life sciences was a key resource for the rest of NASA as it pursued a wider array of cooperative projects with the Russian space agency.

NASA headquarters, however, showed no inclination to squeeze out a peace dividend from the NASA budget. Plans for a Moon colony and a human mission to Mars were abandoned slower than the growing realization that the technology was too premature to do either safely or cheaply. Congress grew more impatient as NASA let the International

Space Station, the key cooperative project, soak up any funding liberated from NASA's defense-oriented projects. On 12 March 1992, Bush made a surprise announcement—that he had nominated Daniel Goldin to replace Richard Truly, whom he had asked to resign as NASA administrator.

Goldin was a vice president and general manager of the TRW Inc. Space and Technology Group in Manhattan Beach, California, which specialized in commercial, spy and early-warning satellites. During Goldin's five year tenure in that group, TRW had built thirteen such spacecraft—for the tracking and data relay satellite network, the Air Force defense support program, and the Brilliant Pebbles and Brilliant Eyes projects of the Strategic Defense Initiative Office. For NASA, TRW had built the Compton Gamma Ray Observatory and parts of the Advanced X-ray Astrophysics Facility. TRW won NASA's 1990 Goddard award for quality and productivity and was a finalist for the George M. Low trophy for excellence. Those who bought spacecraft from TRW knew Goldin as a very capable manager. Those in space policy knew nothing about him.

Goldin's early pronouncements showed him supportive of a smaller International Space Station, a human landing on Mars, and reliable operation of the shuttle. But mostly, he talked about applying an industrial perspective to shake up NASA. "He's a faster, cheaper, better kind of guy," said a Bush administration official. "He's obviously outside the NASA culture."[1]

Daniel Goldin, NASA Administrator

"My challenge," Goldin proclaimed in his first address to NASA employees, "is to convince you that you can do more, do it a little better, do it for less, if we use more innovative management techniques and if we fully utilize the individual capabilities of each and every NASA employee." Goldin also voiced, Ames people noted, distaste for how he perceived NASA's recent work in aeronautics: "We have to perform world class aeronautics research. Not leave it on the back-burners, not enjoy all the fun we're having writing TRs and TNs [technical reports and technical notes], but what we have is an obligation for America. The American aeronautics industry is counting on us and let's ask ourselves, have we really lived up to the expectations of American aeronautics?"[2] He was obviously a man of extraordinary energy, different views and, Ames people soon discovered, of strong personality.

Not the passage of time, nor changes of heart, nor the growing respect for Goldin's leadership—nothing softens the horror when Ames people tell the story of Goldin's first visit to Ames. There is no videotape that recorded what actually happened, so stories are told. Articles criticizing Goldin's intentions had just appeared in Bay Area newspapers and Goldin, one Ames manager remarked, "seemed to show up loaded for bear."[3] Rather than listen to welcoming speeches, he counted the number of women and minorities in a photograph of Ames executives, then made pointed comments about how few he found. Goldin challenged those he happened upon to defend their programs. People hid their name badges. In a meeting in the director's conference room, Goldin sent to the perimeter all those sitting around the table—mostly senior white males—and asked those sitting in perimeter chairs to take their place. Then Goldin heckled director Dale Compton as he reviewed Ames' strengths and goals, until Compton walked silently from the room, halfway through his presentation, to compose himself. Only then did Goldin's wrath subside.

Goldin himself has turned philosophical about how NASA people reacted to the force of his personality. Goldin's visit, in fact, foreshadowed that he really would push for a diverse workplace, for opening up NASA facilities to scientists outside the usual clubs, for imposing total quality management, and for tightening the NASA organization. But clearly, there was more to his displeasure with Ames.

NASA headquarters sent a surprise security review team that descended upon Ames on the evening of 31 July 1992. They sealed buildings, changed locks, searched file cabinets,

Atmosphere of Freedom

Aerial view of Moffett Federal Airfield in 1995, looking straight on the main runway. In the background is the San Jose International Airport.

took computers, interrogated hundreds of scientists, and sent ten researchers home on administrative leave. Only Compton was told, the day before, who they were, what they were looking for, and what prompted the raid. The team pointedly asked everyone about "management's judgment" on technology transfer matters.[4] Rumors circulated that they targeted scientists of Asian descent, especially those in the aerophysics directorate. In the end, the team discovered nothing illegal, and Ames altered some minor security procedures. But some good people decided to quit, and the Center was left with deepened concerns about the attitudes toward Ames that prevailed in NASA headquarters.

Whenever Goldin talked of Ames he used the word "revitalize," which Ames people considered better than "shut down." During the summer of 1992, as Bill Clinton made gains in the polls, Ames people thought a change in administration might remove Dan Goldin from their list of worries. But Albert Gore, as senator from Tennessee, chaired the committee that oversaw NASA matters and liked what he saw in Goldin. When Gore became Vice President, he asked Goldin to stay on.

Moffett Field, Quality, and Cultural Climate

Compton won the next round of tensions between Goldin and Ames—over the reconfiguration of Moffett Field. The Navy had managed Moffett Field since 1931, except from October 1935 (following the crash of the dirigible Macon) to April 1942 when it was run by the Army Air Corps. In the 1950s, the Navy based supersonic fighters there until the community objected to the noise. In 1962, propeller-driven P-3 Orions arrived on base to fly patrols over the Pacific in search of Soviet submarines. With the collapse of the Soviet Union in 1990, the Navy said it no longer needed Moffett Field. The Base Realignment and Closure Commission (BRAC) agreed.

The Bay Area congressional delegation, led by Norman Mineta, a San Jose Democrat who chaired the Congressional Space Caucus, stepped into the fray. They convinced the BRAC that, even if the Navy left, Moffett should remain a federal

Burrowing owls and least terns keep watch over Moffett Field wetlands.

National Full-Scale Aerodynamic Complex (NFAC) wind tunnels and the Numerical Aerospace Simulation (NAS) building from across a springtime field of mustard and wild flowers.

airfield. Efforts in 1990 to declare fifty acres at Moffett as protected wetlands, and to chart the presence of protected species like the burrowing owl, least tern, and peregrine falcon limited other developments at the field. In the October 1991 recommendations approved by Congress and the president, the BRAC said that NASA, as the next biggest resident agency, should become Moffett's custodian. The Navy had subsidized Moffett operations at $6 million per year, a cost NASA then would have to include in its budget unless it found other ways to generate revenues from field operations. Yet NASA administrator Richard Truly understood the opportunities for Ames. Goldin inherited a decision, however, that was not initially in line with his change agenda. NASA headquarters was already planning to trim Ames' flight operations. Furthermore, if Congress ever imposed a BRAC-type process on NASA, headquarters presumably would want nothing to get in its way of shutting down Ames. Compton and his executive staff understood this, marshalled the substantial goodwill toward Ames from its local community, and wrested control of the property on which Ames sat. Not until 23 December 1992, in a subdued signing ceremony at Ames, did Goldin concede that NASA would step up as custodian agency when the Navy officially decommissioned its station in July 1994.

"Finding a Cure for Cancer," a fanciful depiction by Clayton Pond.

"Over the past five years in my prior job, I've become a true believer in the value of total quality management," said Goldin. "I believe deeply that if you can't measure it you can't manage it, and I intend to bring this philosophy to NASA."[5] Total Quality Management (TQM) was like a confusing new language. Throughout the 1970s, headquarters had asked Ames to undertake consultant-driven reviews and exercises—like quality circles—to make itself more efficient, and it was entirely Goldin's prerogative to impose the latest fashion in organizational improvement. But TQM was confusing. It demanded a focus on the "customer." "The space program doesn't belong to us," Goldin would say. "It belongs to the American people. They are our customers."[6] Lots of NASA people did not find that definition specific enough to clarify how they would use all the statistics and acronyms TQM demanded. But Ames people tried.

Jana Coleman addresses the Ames all-hands meeting on 16 July 1992 to explain the process of total quality management.

Compton called an all-hands meeting in July 1992 on the Ames flight line to say Ames would start implementing TQM with a year of education and training. Meanwhile a quality improvement team, chaired by Jana Coleman and Robert Rosen, worked with continuous-improvement consultants Philip C. Crosby, Inc., and wrote a report on the whole TQM process. In April 1993, Ames posted everywhere its carefully worded quality statement. Ames' management council approved the report in February 1993, and set about forming process action teams (PATs) to reduce the costs of non-conformance (CNC). Ames created a culture that naturally supported spontaneous quality teams and continuous improvement. Throughout the Center, teams defined their customers, used flow charting and process measurements, tore apart then rebuilt all their procedures, and began to report savings in costs and time. For example, in late 1993, the Unitary 11 foot transonic tunnel applied a TQM approach to test

Ronald McNair Intermediate School in East Palo Alto has long hosted mentors from NASA Ames. Dale Compton visited in 1991 to encourage the students to prepare for careers in science and engineering.

runs for the Navy's A/F-X competition by four contractor teams. By reviewing their procedures and listening to their customers, the tunnel group doubled the expected number of successful runs. Ames announced a $2 million investment in process infrastructure—like electronic forms and purchasing, computer peripherals, and a charge-back system for technical support—that helped all teams improve their processes. Ames made good progress, even though the Crosby literature trumpeted that continuous improvement is a cultural process that takes five to seven years to change—"so don't let impatience cloud your view of progress."[7] Ames undertook the Malcolm Baldridge Self-Assessment in the fall of 1993—less than eighteen months after starting TQM—because of a Clinton administration initiative to "reinvent government." The survey showed that, even though Ames people thought their work was very high quality, they knew little about Ames' formal quality process. Ames lagged well behind all other organizations actively implementing TQM.[8] Ames management, presumably, had not become true believers in TQM.

Another cultural review further widened the chasm between Ames management and NASA headquarters. In July 1992, Ames was visited by a NASA-wide Cultural Climate and Practices Review Team, led by General Elmer T. Brooks, deputy associate administrator for agency programs. The team gave Ames a glowing report, calling it "the best" of all NASA Centers. Ames employed higher percentages of underrepresented groups than any other NASA Center; the Ames Multi-Cultural Leadership Council was a model for other Centers; participation was strong in the Equal Opportunity Advisory Groups—African American, Asian American and Pacific Islander,

Ames research contributed to the aerodynamics and thermal protection systems of all single-stage-to-orbit spacecraft under development in the late 1990s: (left to right) the Rockwell wing body; the Boeing/Bell vertical landing configuration, and the Lockheed Martin lifting body.

Michael Marlaire, chief of Ames' external affairs office, discusses plans for Moffett Field at a local town meeting.

Disabled, Hispanic, Women and Native American; Ames won NASA's Equal Opportunity Trophy in three of the past nine years; and Ames' entire work force felt challenged and satisfied with their work.

However, there were problem areas. The percentages of minorities employed were lower than in the culturally diverse Bay Area as a whole. African Americans were especially underrepresented, indicating that Ames had failed to reach into the local community. Ames tended to hire experienced researchers rather than those fresh out of co-op programs. Any mentoring was too informal, and career development was haphazard. Higher wages in local industry made it tough for Ames to retain the leaders it did develop. Of forty top managers, only one was a woman and only two were minority males. Minorities and women perceived the senior executive service as a white male preserve. In fact, the Brooks team declared that all problems were caused by upper management. Despite being the best in NASA in affirmative action, the Brooks team reported, "everyone is looking to the Center Director for proactive leadership."[9]

Then, in October 1993, Congress pulled funding for the SETI (Search for Extraterrestrial Intelligence) program that Ames had nurtured for two decades and that had stirred up enormous scientific excitement around NASA. Some Ames staff felt that Goldin failed to stand up to congressional doubts, and sacrificed SETI to secure funding for the space station and for programs at other Centers. Goldin later said that NASA would focus instead on the far more promising search for dumb, organic life in the universe by developing the discipline of astrobiology. Eight civil servants and fifty contractors were affected by the $12 million cut. As other Ames projects were cut, and as Ames prepared for many years of flat or declining budgets, Ames opened a career-transition office to move its work force into a booming Silicon Valley economy hungry for such technical skills.

Compton and Vic Peterson, Ames' deputy director, increasingly felt that, as the lightning rods for some unarticulated displeasure from NASA headquarters, the best thing they could do for their Center was to retire. On 22 November 1993, both Compton—after 36 years of government service—and Peterson—after 37 years of government service—took the retirement they had earned. In declining to speculate on what his successor might consider Ames' major goals and challenges, Compton replied: "The long term goals of this Center have survived many directors."[10]

Ken K. Munechika

Ken K. Munechika, Director of Ames Research Center from 1994 to 1996.

On 17 January 1994 Ken K. Munechika became director of Ames. Munechika was raised in Hawaii and earned a doctorate in educational administration from the University of Southern California. He had a distinguished career in the U.S. Air Force. He started as a navigator, flew 200 combat missions in Southeast Asia, moved into training as a professor of aerospace studies, then served as chief of satellite operations to recover space capsules deorbited from space. In July 1981 he moved to Sunnyvale to command the Air Force Satellite Control Facility (later renamed Onizuka Air Force Station), where he directed contractor teams in launch operations of more than fifty defense satellites, and all the defense payloads launched by NASA's Space Shuttle. He was also responsible for planning and budgeting a global network of satellite tracking stations. He retired in June 1989 to become executive director of the Office of Space Industry for the state of Hawaii (where he would return after being reassigned from Ames).

Munechika asked William E. Dean to serve as his deputy director. Dean, too, was a newcomer to Ames, having arrived in August 1991 as special assistant for institutional management. Prior to that, Dean served as president of Acurex Corporation of Mountain View, a privately held supplier of control and electronics equipment. Before then, from 1962 to 1981, Dean worked for Rockwell International, serving as group vice president responsible for the Global Positioning Satellite and the operational phase of NASA's space shuttle program. Compton had hired Dean to infuse business-like thinking into Ames, and Munechika asked him to stay on.

Ames inscribed its name, again, on NASA's Equal Employment Opportunity Trophy in 1993.

Though he had spent his entire career managing the highest technology in the Air Force arsenal, Munechika was the first to admit he was no scientist. His first priority was addressing the lingering factionalism from the Cultural Climate and Practices Plan. "Since aeronautics and space are for everybody," Munechika wrote, "I want Ames to look like America and the community we represent....Ames must have a work environment where everyone feels empowered, included, valued, and respected."[11] Jana Coleman was named to lead the newly created Center operations directorate, the first woman to head a directorate at Ames. Ames attracted good people, who in turn attracted good people, who then saw Ames as a great place to build their careers. This was a key part of Ames' success, so Ames people addressed their diversity with seriousness.

Ames people also put more vigor into their outreach efforts. Every year for two weeks thousands of students gathered for the JASON project to explore, through telepresence, the scientific mysteries of our Earth. Ames formed a docent corps to staff the Ames Aerospace Encounter, the Ames Visitor Center, and the Ames Teacher Resource Center. NASA distributed internet kits to area schools, and engineers volunteered to share with students the excitement of their work. Ames expanded its relationship with the National Hispanic University (which began early in 1993 with a space sciences program and would culminate in an historic collaborative agreement in October 1997). Interns and research fellows came from a wider variety of schools. Space Camp California opened just outside Ames' main gate.

With Munechika to introduce them, headquarters staff showed up more regularly at Ames, praising its revitalization efforts. Many of the significant events and program activities that would follow—like the Zero Base Review, the information technology Center of Excellence, the Astrobiology Institute, Lunar Prospector, the SOFIA restart, and the absorption of Moffett Naval Air Station—were all started in a fairly short period of time during

Ames tracking and telemetry station at Crows Landing auxiliary airfield.

Munechika's tenure as director. Yet bolstered morale and coalescence of support from the external community only served to brace Ames people for program adjustments and structural changes still to come. The darkening funding picture and Goldin's agenda for change set the challenges for Munechika's period of leadership. The same day Goldin announced Munechika's appointment, he also announced the appointment of three other Center directors (two of whom, like Munechika, would be gone within three years). He further announced that, in March 1994, after twelve years as part of Ames, Dryden would become an independent Center. Ames management expected that, as Dryden asserted itself in NASA planning, programs and people would be shifted there from Ames.

Headquarters let Ames staff know that Moffett Field was their burden to bear. Countless details were ironed out in advance of the transfer, all coordinated by Michael Falarski and Annette Rodrigues of the NASA-Moffett Field development project. But change appeared gradually—access guidelines were redefined, security guards wore different uniforms, the Navy's P-3 Orions left, the Navy began environmental remediation, and historic preservationists surveyed the architecture. In 1993, NASA took control of the small naval airfield at Crows Landing in Stanislaus County,

MD-900 helicopter in hover mode during noise-abatement tests.

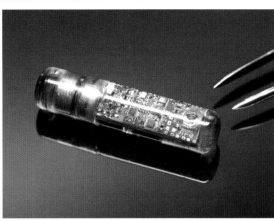

Advanced telemetry devices such as this pill transmitter can monitor fetal health in the mother's womb.

Decommissioning ceremony on 1 July 1994, marking the transfer of Moffett Federal Airfield from the U.S. Navy to NASA.

Toni Ortega and Lisa Hunter evaluate a mock-up of the Space Station centrifuge facility in July 1994.

which Navy pilots had used for P-3 training flights and which NASA would use for low-speed flight research. The Onizuka Air Force Station took over the military housing that Navy families vacated. On 1 July 1994, while a Navy blimp and a P-3 Orion flew overhead, a 21 gun salute and taps sounded as Navy officers lowered their flags. "From Lighter than Air, to Faster than Sound, to Outer Space:" that's how the Navy commander described the changes seen at the Moffett Field Naval Air Station. NASA renamed it Moffett Federal Airfield to reflect its new organizational flexibility. It now could serve a wider array of tenants and customers—the Naval Air Reserve Santa Clara, the Army Reserve, the California Air National Guard, other governmental agencies like the U.S. Post Office and the Federal Emergency Management Agency, and private firms executing government contracts. Then Ames people started planning to make something new and exciting from their enormous facility and opportunity.

"Solid Smoke" aerogel insulations developed for the Space Shuttle have found uses on Earth such as insulators for refrigerators, furnaces and automobiles.

Ames started by assessing community needs in the adjacent cities of Mountain View and Sunnyvale and in the Silicon Valley region. San Jose International Airport was congested, with any expansion limited by its proximity to downtown and its location amid residential neighborhoods. Moffett Field offered a superb airfield—twin runways 9,200 feet and 8,900 feet long, ample tarmacs, three very large hangars, aircraft fuel and wash facilities, and more than seventy structures for aircraft operations. It had round-the-clock crash and rescue service, sixteen hour air traffic control, instrument landing equipment, world-class communication links, and easy access to Highway 101. What it lacked was air traffic, so Ames facility managers suggested using the airfield for business and freight flights. Specifically, the San Jose airport could no longer fit in jumbo jets ferrying electronics back and forth from Asia. Furthermore, Bill Dean, Ames' deputy director and the person most responsible for base planning, thought that Ames should keep the airfield as the Navy left it. Like so many others, he thought that some day soon Russian submarines would again patrol the Pacific and the Navy would return its P-3 Orions. Converting Moffett Field into an air cargo base best kept it in mobilization shape, so that was the plan he proposed.

But local residents had gotten used to quiet (even though the P-3 and C-130 flights were never very noisy). Rather than decide themselves, the Mountain

Chris McKay explores for life in the harsh environment of Dry Valley, Antarctica.

A Center Reborn: 1990 – 1999

The Slender Hypersonic Aerothermodynamic Research probe (SHARP-L1) is an example of a revolutionary lifting body concept.

View and Sunnyvale city councils asked for a nonbinding vote on the plan to make Moffett Field a freight airport. Voters advised against the plan, Munechika respected the vote, and Ames was left to devise another plan while shouldering the costs of running the base. Though the derailing of momentum behind the Moffett Field plan was a loss, far more significant losses came in the wake of NASA's Zero Base Review.

Zero Base Review

Goldin arrived at NASA proclaiming that NASA was bloated. He imposed a new type of discipline to NASA's budget process and, in time for the fiscal 1994 appropriations, submitted a budget that reduced NASA's five year budget by $15 billion. Two years later, by cancelling programs and redesigning the International Space Station, he reduced NASA's long-range budget by thirty percent. He called this process "a fiscal declaration of independence from the old way of doing business." But by 1995, when Congress asked NASA to absorb an additional $5 billion in cuts from its $14 billion budget, starting in 1997, Goldin realized that the loss of more research programs would jeopardize NASA's leadership in aerospace technology. So in response to the Clinton administration's call for a National Performance Review, instead of cutting programs Goldin focused on streamlining NASA's infrastructure through a Zero Base Review (ZBR).

Rather than starting with last year's budget to develop the next, zero base budgeting means starting from zero every year, and asking whether each program is essential to an agency's core missions. This was different from the national laboratory review of 1992, which focused mostly on eliminating duplication of functions. A headquarters "red team" visited in 1994 and asked Ames people to ponder the prospect of being shut down. The preliminary ZBR white paper of April 1995, drafted by NASA deputy chief of spaceflight Richard Wisnieski, translated this vague recommendation into a specific budget planning document. Nancy Bingham, the Ames manager on whose desk the faxed ZBR draft landed, called it "inflammatory."[12] It presented numbers that dropped Ames civil servant cadre from 1,678 to below 1,000 within five years—below the point of viability. Aerospace facilities would be transferred to Dryden, and the space station centrifuge would go to Johnson Space Center. What remained of Ames could then easily be shunted into a GOCO—a government owned, contractor operated facility. Ames had in the past confronted efforts to shut it down—in 1969 at the start of Hans Mark's tenure and during the 1976 reductions in force before he left. The draft ZBR white paper made it most clear—in dollars and headcounts—that if people in Washington wanted to rebuild NASA from scratch, they would rebuild it without Ames.

To stave off the threat that the entire Center would be shut down immediately, Ames mobilized support within the community, among California legislators and Ames' friends in Washington. Congressman Norm Mineta protested that the people of Ames "are too valuable to be left to the underestimation of NASA bureaucrats in Washington."[13] With the small amount of time they won, they dove head first into the challenge of zero-base thinking. NASA headquarters had started by defining its five strategic enterprises—mission to planet Earth, aeronautics, space science, space technology, and human exploration and development of space. They intended to declare each Center a center of excellence in some area to help all of NASA execute those missions. Each Center would take on lead center programs, and administrative functions would be consolidated agencywide. Deciding which Centers should execute a mission and which were "overlap" got intensely political.

Many at Ames believed their Center did not fare well in the grab for assignments. Ames lost its leadership in Earth sciences to Goddard, in biomedical sciences to Johnson, in space technology to Marshall Space Flight Center, and in planetary sciences to the Jet Propulsion Laboratory. Significantly, Ames lost its leadership in aerodynamics and airframes to Langley, and Langley would also manage Ames' tunnels and simulators, which were mostly staffed by contractors but made up sixty percent of Ames' budget. Ames faithfully eliminated programs declared redundant, and executed its plan for 35 percent attrition during 1996: buyouts reduced the number of civil servants by 300, layoffs almost halved the number of contractor personnel to 1,400. Most importantly, Ames lost its aircraft to Dryden.

In December 1990 NASA headquarters had appointed long-time Dryden researcher Kenneth Szalai to the position of "Director" of the Ames-Dryden Flight Research Facility. Marty Knutson, who had managed the facility for five years and guided Szalai's development as a manager, returned home to Ames. Goldin visited Dryden in September 1992 and announced that "the right stuff" still lived there and, indeed,

Szalai proved adept at bringing new projects to Dryden—from industry as well as from other NASA Centers. By March 1994, after thirteen years of direction from Ames, Dryden again became an independent NASA Center. In a note to Ames employees, Szalai wrote "Many professional associations and friendships were developed and I intend to work hard to sustain these….Please consider Dryden as your flight research center, too."[14] Then, on 19 May 1995, NASA announced that for cost savings every aircraft in the NASA fleet—operational as well as experimental—would be consolidated at Dryden.[15] Ames had the most to lose. Of the seventy aircraft in NASA's fleet, Ames then serviced twelve—three ER-2s, one DC-8, one C-130, one Learjet, one C-141 and five helicopters. Moving the airborne science airplanes provoked the most controversy. Ames management argued that these airborne laboratories relied on input from an active scientific community simply not found in California's high desert, and that they used equipment made in Silicon Valley. "This consolidation could mean the end of valuable environmental programs," wrote California Congresswoman Anna Eshoo, "I'm also concerned NASA is fudging its fiscal homework on the consolidation plan.

Its numbers are incomplete and its economic justifications are questionable."[16] The flight operations branch, the first branch established at Ames, was disbanded. Some support staff moved with the aircraft; some retired, like long-time flight operations chief Martin Knutson and pilot Gordon Hardy; most took new assignments at Ames. In November 1997 the last Ames aircraft flew off to Dryden. A disconcerting quiet hung over the Ames hangars. Researchers at Ames who had dedicated their careers to improving aircraft and who wanted to see them in flight, now had to shuttle south to the desert on a little commuter airplane.

Amid all these program losses, however, Ames had constructed a bold new strategy. Ames' response to the ZBR fell on the shoulders of a mid-career group of technical leaders—most of whom had hired into Ames during the 1970s and had honed their advocacy skills in the strategy and tactics committee meetings called by Bill Ballhaus and Dale Compton. Despite the mandate of zero-based thinking, they refused to believe that Ames had no history. They knew the people there, how fluidly they worked together, and how ingeniously they used the research tools available. But Ames management had not done a good job marketing these capabilities. Coordinating efforts from the Ames headquarters building, Nancy Bingham, Bill Berry, Mike Marlaire, Scott Hubbard and George Kidwell pulled together comments from their colleagues around Ames, and gradually a strategic response emerged. Ames polished this story by talking to community leaders, to the Bay Area Economic Forum, and the local press. The final NASA ZBR white paper of May 1996 showed Ames' headcount at 1,300 and that Ames would lead NASA in information technology, astrobiology, and aviation system safety and capacity.

Ames' response to the ZBR marked its rebirth. In the same way that so many scientists and engineers had reinvented themselves to address new national needs,

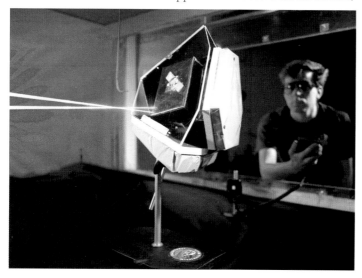

Mars Pathfinder model being installed by Chris Cooper in the Mars wind tunnel at Ames.

Henry McDonald, Director of Ames Research Center since 1996.

by the end of the ZBR exercise the Center had also redefined itself. It coincided with the arrival of a new leader, who understood Ames' past and its future.

HENRY McDONALD

Harry McDonald remembers that when he first met Goldin, Goldin said that he "gave Ames one plum assignment—to become a center of excellence in information technology—and that Ames hadn't executed it well."[17] Munechika's plans for the newly created information systems directorate were largely derailed when David Cooper, the information system director he appointed to replace Henry Lum, left and many of his staff left with him. Consolidating all of NASA's computing and communication systems should have shown a savings of 1,200 positions nationwide, but the systems were still burdened by disorganization and redundancy. More and more NASA projects revolved around imaging, robotics, data crunching and internetworking, and NASA people had a hard time finding the expertise they needed. Moreover, Ames was stagnant. Since becoming a part of NASA, Ames had always had about seven percent of NASA staff and about five percent of its budget. If Ames expected to grow it had to take a bold stance. Thus charged with implementing Ames' information technology mission, McDonald arrived as Ames director on 4 March 1996.

A native of Scotland with a doctorate from the University of Glasgow, McDonald had spent the previous five years as professor and assistant director of computational sciences in the Applied Research Laboratory at Pennsylvania State University. Before that, McDonald was president of Scientific Research Associates, Inc., of Glastonbury, Connecticut, a company he founded in 1976 to do contract research in computational physics and gas dynamics. The state of Connecticut awarded McDonald its small businessman of the year award for high technology because of a ventilator he invented and developed. And before that he worked as a research engineer for British Aerospace and then for United

Yuri Gawdiak displays a personal satellite assistant prototyped at Ames in 1999 for future spaceflights. The PSA is an autonomous robot based on Ames' work in telepresence, artificial intelligence, and research in microgravity. It hovers behind an astronaut, monitors environmental conditions, collects data from onboard experiments, provides video conferencing, and otherwise serves as a wireless link between the astronauts and ground crews.

Technologies where, along with colleagues at Ames, he developed the linearized block implicit methods for solving compressible flow equations. McDonald joined Ames on an interpersonnel agreement (IPA) that allowed him to keep his university tenure, and he kept his house in Glastonbury, where his wife had her medical practice.

McDonald was an expert in computational aerodynamics, and people around Ames knew and respected his work. As his deputy director, he named William Berry who had built a strong reputation for management in the space and life sciences side of Ames. McDonald also brought in new managers from the outside—like Robert "Jack" Hansen as deputy director of research, and Steve Zornetzer as director of information sciences and technology. He also invited back an old hand as his advisor, John Boyd.

Intellectually, McDonald understood the entire range of work at the Center and could thus represent it effectively outside. He tempered what many perceived as the traditional arrogance around "Ames University." (While speaking at Ames, General Sam Armstrong once opined that NASA was a place, "where a direct order is seen as an open invitation to debate.") McDonald tapped in to the desire of Ames researchers to embrace change rather than fight it, and to constantly reinvent themselves by applying their skills to new challenges. Most important, McDonald focused Ames on implementing the strategic opportunities posed by the Zero Base Review.

AME-2 airplane, designed to be unfolded and flown through the atmosphere of Mars, in flight demonstrations at Moffett Field.

Center of Excellence for Information Technology

"The future of NASA lies in information technology and information systems," proclaimed Goldin in May 1996 in a ceremony designating Ames as the NASA Center of

A Center Reborn: 1990 – 1999

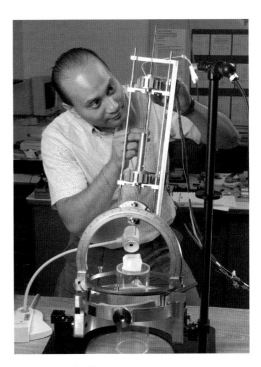

Michael Guerrero works with the robotic neurosurgery apparatus, a computer controlled motor driven smart probe with a multisensory tip.

Excellence for Information Technology (CoE-IT). The CoE-IT developed rapidly, directed by Jack Hansen and then Kenneth Ford, as the center of a virtual corporation that linked NASA Centers, industry, and academia into tight-knit teams. These teams developed "enabling technologies" in modeling, database management, smart sensors, human-computer interaction, and supercomputing and networking. These enabling technologies then supported key areas of NASA's missions — integrating aerospace design teams, networking data to improve simulations, improving efficiency in aviation operations, and developing autonomous probes to make space exploration more frequent, reliable and scientifically intense. Ames led NASA efforts to incorporate advanced information technologies into all NASA space and aeronautics programs in support of faster, better, cheaper programs.

The CoE-IT consolidated resources at Ames and from around NASA. Ames became NASA's lead center for supercomputer consolidation, overseen by the Consolidated Supercomputing Management Office (CoSMO). Consolidation began with an inventory of all of NASA's supercomputers—including the central computer facilities, the NAS facility, and the test beds in the high performance computing and communications program—and identified forty systems with a total purchase price of $300 million. Consolidation continued as Ames matched the right computer to the right job within NASA's enterprises. For example, the NAS and the computational chemistry branch together pioneered new ways of designing nanotechnology—machines built at the molecular level.

From there, Ames information technologists set about building NASA's information power grid, a system of interlinkages

A surgical robot, designed to probe breast tumors and to be controlled over the internet, developed by Robert Mah of the Ames bioinformatics center.

Christine Falsetti and John Griffin review Ames contributions to building the next-generation internet.

that enabled distributed supercomputing to support the entire range of NASA research and decision-making. The tools they developed for the seamless integration of computing and data archiving also underlay NASA's contributions to NREN, the national research and education network. Ames' CoE-IT was also named to represent NASA as the federal government invested $300 million to build the next-generation internet. Christine Falsetti led a group of thirty Ames people designing and integrating new network technologies that would allow data to flow a thousand times faster than in 1997.

One Ames effort integrated into the CoE-IT was the NASA center for bioinformatics, opened in August 1991 with a dazzling display in the Ames auditorium by Muriel Ross. A biologist specializing in the neural networks around the vestibular system, Ross joined Ames in 1986 for access to its supercomputing infrastructure. She suspected, and later experiments would confirm, that exposure to microgravity caused the inner ear to add new nerve cells. She also suspected, rightly, that this rewiring could only be accurately depicted in three-dimensional models. Reconstructing the architecture and physiology of this expansive neural network was painstaking work.

Muriel Ross and Rei Cheng looking over the immersive work bench in 1996 showing a skull for facial reconstructive surgery.

Daryl Rasmussen working in Ames' Mars map laboratory. Ames provided much of the visualization technology that helped Mars Pathfinder grab headlines around the world.

So Ross worked with programmers in the NAS to devise a technology for reconstructing serial sections of a rat's vestibular system into a three-dimensional computer model. Ames' artificial intelligence experts explored this model for clues about building neural networks with computers. Ames experts in virtual reality, led by Glenn Meyers of Sterling Software, bought a prototype virtual boom from Fakespace Corporation and linked it with Silicon Graphics workstations to project reconstructed images into the first immersive workbench. There, surgeons could rehearse difficult procedures before an operation.

The next step for the bioinformatics center was to build collaborative networks with other NASA centers using emergent Silicon Valley networking technology. Stanford University Medical Center was first, followed by the Cleveland Clinic Foundation, then the Salinas Medical Center, and the Navajo Nation. With each new collaborating clinic—each more distant and less sophisticated in computing—Ames tested technologies for doing remote medicine, preparing for when astronauts many days distant on the space station might need to respond to medical emergencies. In the meantime, the center is a national resource that allows investigators to apply advanced computer technology to the study of biological systems. When challenged to apply its skills to a national initiative in women's health, the Ames center for bioinformatics developed the ROSS software (for reconstruction of serial sections) to provide very precise three-dimensional images of breast cancer tumors.

Virtual reality applied to neuroscience. Rei Cheng of the Ames bioinformatics center shows a model made to study motion sickness and balance in the inner ear.

Ames' telepresence control room, in practice for robotic exploration of distant planets.

Ames made telepresence the key to NASA plans for planetary exploration. In 1990, the Ames space instrumentation and studies branch, led by Scott Hubbard, developed mission plans for the Mars environmental survey (MESUR). The plan was to build a global network of sixteen landers around the Martian surface—each capable of atmospheric analysis on the way down and, once on the surface, of performing meteorology, seismology, surface imaging, and soil chemistry measurements. Because the network could grow over several years, the annual costs would be small and the landers could be improved to optimize the scientific return. With the data, NASA could pick the best spot to land a human mission to Mars. However, in November 1991, NASA headquarters transferred MESUR to JPL, where it was trying to centralize work in planetary exploration. JPL invented the idea of a single MESUR lander, renamed it and developed it into Mars Pathfinder, which roved across the Martian landscape in July 1997.

Ames continued developing the technology to support robotic missions to Mars after MESUR had been moved. In January 1992, Geoffrey Briggs was appointed scientific director of Ames' new center for Mars exploration (CMEX). Since the Viking missions of the mid-1970s, Ames maintained a world-class group of scientists specializing in Martian studies across a broad spectrum. CMEX brought all of this expertise—especially expertise in robotic spacecraft and data

At the distant end of an internet line a physician, at the Cleveland Clinic Foundation working in partnership with Ames to develop telemedicine technologies, checks an echocardiogram displayed in real time.

A Center Reborn: 1990 – 1999

Ames is wired into the heart of NASA's information power grid. Al Ross and Richard Andrews in Ames' telecommunications gateway facility.

processing—to bear on key questions in the geographical and atmospheric evolution of Mars. Ames paleontologist Jack Farmer and exobiologist Christopher McKay led studies on the best strategies and locations to search for life on Mars.

"Antarctica is the most Mars-like environment on Earth," said Carol Stoker of the Ames telepresence technology project. "We're taking this technology to a hostile environment to conduct research that has direct application to NASA's goal of exploring Mars."[18] In December 1992, Stoker and Dale Anderson tested telepresence technology on minisubmarines exploring the sediments under the permanent ice covering Antarctic lakes. The next Antarctic summer they returned with a rover with stereoscopic vision, not only so they could generate a three-dimensional terrain model of the McMurdo Sound but also so the teleoperator had depth perception to better collect samples with the rover's robotic arm. Back at Ames, Butler Hine controlled it using a teleoperations headset developed by Ames' intelligent mechanisms group. They were linked via a powerful satellite and internet connection put together by Mark Leon and the NASA science internet engineering team.

Ames' artificial intelligence software guided the Deep Space 1 spacecraft, launched on 24 October 1998. It was the first mission under NASA's new Millenium program to test in spaceflight the many innovative technologies that will lead to truly "smart" spacecraft. One new technology was Ames' AutoNav remote agent that made the spacecraft capable of independent decision-making so that it relied less on tracking and remote control from the ground. In May 1999, for the first time, an artificial intelligence program was given primary control of a spacecraft. On 28 July 1999, after getting a brief instruction to flyby the asteroid 9969 Braille, the DS-1 remote agent evaluated the state of the aircraft, planned the best path by which to get there, and executed a flyby no more than ten miles from the asteroid.

The Ames CoE-IT, managerially, was increasingly integrated into the Ames information science and technology division as ways of applying this expertise became a more normal part of Ames' operations. Ames assumed oversight of the NASA Independent Verification and Validation Facility in Fairmount, West Virginia that independently tests and validates new software for space projects. Ames is applying its skills to test Shuttle avionics software, to make commercial software compatible with proprietary software already used in the Shuttle, and to create an integrated vehicle health management to further expedite Shuttle maintenance. Ames also applies its expertise to help NASA develop aerospace hardware quicker and cheaper, with less technical risk. Integrated design systems, for example, let engineers see and test a system before metal is ever cut. Ames information technologists have systems to translate, in real time, massive amounts of data into visual images and useful information. This has already proved useful in monitoring environmental changes—like fires, hurricanes and ozone holes—from space.

Harry McDonald especially encouraged everyone at Ames to inject intelligent systems and information technology into their work. Revolutionary computing—that is, computing with nanotechnology or neural networks—opens up new opportunities in intelligent flight controls. Ames is developing autonomous systems—essentially an array of sensors, robots, and artificial intelligence systems for non-human space exploration. And Ames information technologists still apply their expertise to solve the logistics and information problems of the airspace system.

Ames has a long tradition of basic research on ways to fight pilot fatigue.

Operation trials of the Ames Center TRACON Automation System installed in 1996 at Dallas/Fort Worth International Airport.

Lead Center Mission in Aerospace Operations Systems

Ames' collaboration with the Federal Aviation Administration (FAA) grew more vibrant in the 1990s. As the federal agency responsible for the national airspace system, the FAA often turns to Ames for technologies to infuse that system with greater safety, efficiency and timeliness. In November 1996, Ames announced a NASA/FAA integrated plan to focus the various facets of Ames' air traffic management research and technology. In June 1997, NASA announced a $450 million aviation system capacity program, with Ames as the lead center to make bold technological leaps forward in air traffic control.

Previously, Ames had focused on human factors in air traffic control and pilot workload. For example, a long running experiment by Curtis Graeber proved that short periods of rest dramatically improved pilot performance during long-haul international flights. Ames' aerospace human factors research division, in October 1993, installed a Boeing 747-400 simulator in its crew vehicle systems research facility (CVSRF). The cockpit simulator was identical to those used to train airline pilots, except that the new displays were reprogrammable and

Heinz Erzberger displays the complicated algorithms he devised as part of Ames' work to improve air traffic safety.

Ames combined its expertise in graphics and air system safety to develop this system for pilots to visualize what's around them even in dense fog or heavy rain.

stocked with equipment for collecting computer, audio and video data. "Our goal is to find ways to improve human capabilities using automation," said CVSRF manager Robert Shiner. Indeed, one of the first experiments evaluated how to replace voice communication between pilot and controller with a digital datalink.[19]

"Modern flight management systems in today's aircraft help pilots do their jobs much better. The CTAS program is about providing the same benefits to air traffic controllers," noted Heinz Erzberger, the Ames research scientist who conceived much of the technology.[20] Ames had been working on traffic control issues since the early 1970s, but they took their research into development with the advent of graphical computers. Programming started in 1991, in consort with the FAA and a team of contractors. In May 1997, Ames released its Center TRACON Automation System (CTAS), a suite of software that generates new types of information to "advise" air traffic controllers. The traffic management advisor picks up aircraft when they are still twenty minutes from landing, and develops an optimum plan for them all to land. The descent advisor graphically depicts the time and

Ames' FutureFlight Central, dedicated in December 1999, is the world's most sophisticated facility for basic research on movement into and around airports.

A Center Reborn: 1990 – 1999

The high bay of the crew vehicle systems research facility: to the front is the Boeing 747-400 cab and to the rear is the advanced cab flight simulator for basic human factors research on pilot workload.

space relationships between incoming aircraft as they converge on an aerial gate forty miles from the airport. The final approach spacing tool lets controllers quickly make corrections to aircraft spacing as they approach the runway.

CTAS quickly proved its value in both time and cost savings at some of America's busiest airports. As quickly as May 1992, Ames installed the simplest version of CTAS at the Stapleton International Airport in Denver, then continued to refine the more complex parts. It passed a major test at the Dallas/Fort Worth International Airport in 1994.

Once aircraft are on the ground, a different set of advisors chime in. The surface movement advisor (SMA) is a simple client-server computer system that collects data so controllers can provide the airlines with automated data on when aircraft will arrive on the ground or at the gate. This relatively simple idea improved gate and crew scheduling, reduced voice traffic over radios, and got aircraft onto the ramp and into the air more quickly. From a go-ahead in March 1994, Ames got a rapid prototype of the SMA working and installed at the Hartsfield Atlanta International Airport in time for the 1996 Olympics. After eighteen months of operations, taxi-time reductions for all aircraft averaged one minute each. Delta Airlines calculated that SMA saves them $50,000 a day in fuel costs alone. The FAA is now exploring ways to make the SMA standard equipment at all airports.

Increasingly, the Ames approach used new information technologies to integrate air-based and ground-based traffic systems. Ames researchers explored unrestricted flight routing, or free flight, which allows more aircraft to safely share airspace under all weather conditions. Ames' advanced air transportation technology branch developed an integrated block-to-block planning service that allows each aircraft to choose its own

From the control room of the crew vehicle systems research facility, Ames experts in human factors study how to improve pilot workloads.

The reconfigurable cockpit interior of the advanced cab flight simulator.

best flightpath, saving minutes of air time per trip and millions of dollars in aggregate.

Another spectacular example of the application of information technology to solve safety issues is the Future Flight Central facility (FFC, though it was first named the surface development test facility). "Surface movement around airports," said Stanton Harke, manager of the Ames program, "is really the bottleneck to making the air transportation system more safe and efficient."[21] Originally conceived to test new versions of Ames' surface movement advisor, the Ames information technology staff saw ways to make something better, faster and cheaper. They used off-the-shelf video and the latest in Silicon Graphics imaging computers to provide a high resolution display with a 360 degree view out the window. Coupled with a sophisticated and changeable console design, for less than $10 million the FFC became the world's most sophisticated test facility for air and ground traffic control simulations. The FFC can be configured to look like the air traffic control tower of any of the world's major airports—both in the arrangement of modular equipment inside the tower and in the view out the window. By reprogramming the display, airport designers can see how well aircraft can move around a proposed airfield, and evaluate any technical innovations, or revise procedures, before concrete is poured.

Ames also completed a system to automatically record and process huge amounts of real-time flight data from new aircraft. "We can detect accident precursors that we didn't know existed," said Richard Keller of his work on the FAA/NASA aviation safety reporting system. Alaska Airlines and United Airlines helped Ames demonstrate the recorder, beginning in 1998, and reported that the data returned could be used to not only improve safety, but also aircraft performance and maintenance scheduling.

Astrobiology Program and Institute

Lake Hoare, Antarctica, and Beryl Spring, Yellowstone National Park.

Martian meteorite.

In addition to its leadership in information technology and air traffic control, Ames accepted the lead center mission in astrobiology—defined as the multidisciplinary study of life in the universe. Astrobiology incorporated the issues early explored as exobiology—the origin of life within the context of evolving planetary systems, and how life evolved, specifically within Earth's harshest environments. Astrobiology also addresses the distribution of life, and how we might search for other biospheres in our solar system. It addresses the destiny of life, how life might adapt to environments beyond Earth, and how life might end as it may have on Mars or Venus. And it includes any scientific approach to these issues—observational, experimental and theoretical.

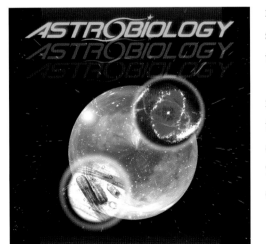

The term "astrobiology," as well as revolutionary plans to pursue it, were sparked to life in the intense pressure and complex chemistry of the primordial zero base review. The future of space science at Ames looked especially bleak, since Ames would likely get no

Assembly of the International Space Station.

new space missions to manage. NASA headquarters had decided that it could support only two centers pursuing space exploration missions—and that Goddard was best established in Earth orbit missions and JPL in planetary missions. NASA chief scientist France Cordova chaired discussions on the role of science within NASA, which were very sensitive to the excellent work done at Ames. She suggested—given the chronically, and now acutely, threatened status of the space, Earth, and life sciences at Ames—that those scientists be privatized outside of Ames in association with a local university.

The idea of a privatized institute, however, hit roadblocks. David Morrison and Scott Hubbard contributed to an agency-wide review, led by Al Diaz of NASA headquarters, of possible approaches to the institute form, but each encountered problems over how to move civil servants into a private institute. Congress balked at passing legislation that eased post-employment restrictions for NASA employees or allowed them to transfer their pensions. Without it, the universities balked at undertaking so big a task as that of integrating an entire research directorate with 600 civil servants and 1,000 support contractors. More important, the institute science plan lacked a cohesive vision. It would be called simply, the Institute for Space Research. In a continuing series of meetings to define a forward-looking agenda for this institute, NASA associate administrator for space science, Wesley Huntress, first suggested the term "astrobiology."

Exciting new scientific announcements in 1996 fueled interest in astrobiology—the discovery of new planets around other stars and hints of fossils in a meteorite from Mars. In August, data from the Ames-managed Galileo probe returned data on Jupiter's climate drivers. The Galileo orbiter returned photos that showed that Jupiter's moon Europa may harbor "warm ice" or even liquid water—both key elements in sustaining life. Goldin saw biology as a

The Space Shuttle orbiter undocking from the space station Mir in April 1996. The Ames life sciences division designed many biological experiments that were transported to Mir's microgravity environment during these encounters.

science with a future, and appreciated all that Ames had done to define the field. He named Ames as NASA's lead center in astrobiology, and tasked it to continue exploring ideas to promote collaboration with larger communities through an institute. The result was a "virtual" astrobiology institute led by Scott Hubbard as interim director.

The NASA Astrobiology Institute (NAI) is in essence creating a new discipline. Ames people created the disciplines of computational chemistry and computational fluid dynamics in the 1970s, driven by the need to theorize the experimental work they had begun. Based on that experience, they took a more deliberate approach to creating astrobiology. A series of astrobiology roadmap conferences identified the holes in the discipline they would need to fill, and established a "virtual" institute that linked universities and research organizations across the United States. In June 1998, the Institute director's office opened at Ames, and accepted competitive proposals to fund research projects. One of the eleven funded projects was from Ames, extending its tradition of research into organic astrochemistry, planetary habitability and early microbial evolution. Plans are to wire this virtual institute into a new and separate building at Ames, where researchers can come to use sophisticated laboratories that most directly support the government's interest in astrobiology—the quarantine and analysis of planetary samples, and mimicking the harsh vacuum, radiation and chemical environments of space. To recognize the major integrative role of astrobiology in Ames' future, in August 1999 McDonald renamed the Ames directorate overseeing all space, earth and life sciences to the astrobiology and space research directorate.

Gabriel Meeker examines her crop in the Ames gravitational biology facility.

McDonald decided that the Institute should be led by a scientist with a world-wide reputation as sterling as what the Institute intended to accomplish. In May 1999 the

Dan Goldin and Harry McDonald introduced Barry Blumberg.

Institute announced that its new director would be Baruch S. Blumberg, who shared the 1976 Nobel prize in physiology and medicine for his work on the origins of infectious diseases that led to the discovery of the hepatitis B virus and discovery of the HBV vaccine. Blumberg would be the first Nobel laureate ever employed by NASA. Goldin turned Blumberg's appointment into an opportunity to make a major address on NASA's vision of exploration, and capped the day by signing an agreement between NASA and SGI, Inc. on a long-term plan to develop new supercomputers. Goldin exclaimed: "It doesn't get much better than this."[22]

Lead Center for Gravitational Biology and Ecology

Likewise, NASA headquarters named Ames its lead center for gravitational biology and ecology to recognize its reputation in life science payloads. Though it appears to be "small science," compared with the "big science" of most other NASA projects, gravitational biology is very management-intensive. Microgravity can only be sustained in space, where it is expensive to send living things. If a space-borne animal is to be sacrificed, every tissue from its body will be studied for microgravity effects. Careful management is needed at every step: to select the experiments from hundreds of proposals; to oversee the very precise construction of habitats and biosensors; to ensure that tissues are carefully prepared and distributed equitably around the

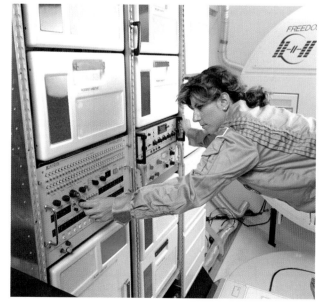

Simulated microgravity test for the Space Station gravitational biology research facility.

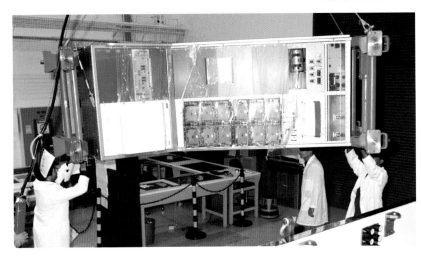

The research animal housing facility being prepared at Ames for Neurolab.

world; to involve every interested biologist in reviewing the data; and to make sure the results are repeatable from flight to flight with very small numbers of subjects.

In the mid-1990s, Ames' work in gravitational biology shifted to the Shuttle/Mir program. These experiments continued the collaboration begun with the Cosmos/Bion missions (which NASA cancelled in 1997) and preceded the biological research to be done on the International Space Station. From June 1995 to January 1998, Ames managed several experiments transferred during the eleven dockings between the Shuttle and the Russian Mir space station. For the first time, a complete life cycle (seed-to-seed) of plants was lived in space. Desert beetles, previously flown on Cosmos/Bion flights, demonstrated the effects of extended space travel on the body's internal clock. Ames researchers swapped tissue cultures with their Russian counterparts, gave the Russians a strain of wheat to grow aboard Mir, and supplied cardiac monitors and bone measuring devices for Mir cosmonauts. Meanwhile, Ames flew a number of experiments collaboratively with the European Space Agency using its Biorack hardware.

Neurolab was NASA's primary contribution to the "Decade of the Brain," as Congress declared the 1990s. The Neurolab mission was launched on 16 April 1998 for 17 days aboard STS 90 and included a variety of experiments to explore neurological and behavioral changes in space. The laboratory contained a

The Biona C—a miniature, computerized blood analyzer developed by Ames' Sensors 2000! program.

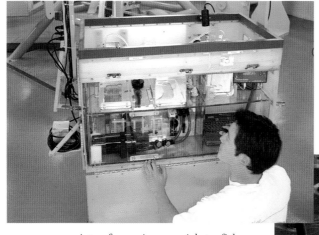

An experiment package to study zebrafish in microgravity is tested in the Ames aquatic centrifuge.

variety of organisms—crickets, fish, mice and rats, as well as monitors for the Shuttle astronauts. Neurolab was one of the most complex missions ever flown in the NASA life sciences. All the non-human experiments—15 of the 26 total experiments—were managed by Ames. In addition, Muriel Ross designed one of the experiments—her third experiment on a Spacelab mission—which led to exciting new reinterpretations of neural plasticity in space. Neurolab was to be the last life sciences mission NASA planned to launch— Shuttle flights would focus on construction of the space station—until the opening of the space station itself.

The space station biological research facility (SSBRF) someday will be the world's first complete laboratory for biological research in microgravity. Ames began designing the SSBRF in the late 1980s as a major scientific module for the International Space Station, though the SSBRF module has been redesigned as often as the space station itself. Meanwhile, the Ames SSBRF project team, led by John Givens, continued to test and perfect the research tools and scientific plans for the facility. A centrifuge measuring 2.5 meters in diameter rotates at selectable rates from 0.01 g forces to 2 g forces, allowing for experiments or experimental controls in artificial gravity. A set of self-contained habitats—with the entire environment remotely monitored from Ames or aboard the space station—will completely support a variety of life forms: rats and mice, insects, plants, small fresh water and marine organisms, avian eggs, and one-celled organisms. A glove box will allow two biologists to perform dissections, transfer samples, and conduct photomicroscopy while keeping the biological samples immobilized and isolated from the rest of the space station. Flash freezers will preserve samples for return to Earth. And a sophisticated data collection system will telemeter data back to scientists at Ames, who will then convey it to university biologists around the world. Ames continues to solicit proposals for experiments from collaborating biologists, so that the experiments run on the SSBRF will study the effects of microgravity on virtually every physiological system. All together, Ames is applying its expertise to enable human exploration of space by understanding a major force—gravity—that differentiates life on Earth from life in space.

Lead Center for Rotorcraft Research and Technology Base

When NASA headquarters transferred other Ames aircraft to Dryden, the Army aeroflightdynamics directorate insisted that its research helicopters should stay at Ames. After several years of negotiation, in July 1997 NASA headquarters signed a directive that Ames would continue to support the Army's rotorcraft airworthiness research using three helicopters. One UH-60 Blackhawk configured as the RASCAL (Rotorcraft Aircrew Systems Concepts Airborne Laboratory) remained as the focus for advanced controls. The NASA/Army rotorcraft division, led by Edwin Aiken, used it to develop programmable, fly-by-wire controls for nap-of-the-earth maneuvering. Another UH-60 Blackhawk was rigged for air loads tests, and an AH-1 Cobra was configured as the Flying Laboratory for Integrated Test and Evaluation (FLITE). In addition, the rotorcraft division made good use of the refurbished wind tunnels. For example, Stephen Jacklin led load and efficiency tests in the 40 by 80 foot wind tunnel of the advanced rotor hub, without hinges and bearings, designed by McDonnell Douglas for its new generation of helicopters.

FASTER, BETTER, CHEAPER

Even in areas where other NASA Centers provided management, Ames has been named leader of specific, important projects. Ames had the history, capability, and people for doing things faster, better, and cheaper. The Pioneer series of space probes, launched in the early 1970s, stand as the best examples of this tradition. This tradition combined well with its ability to craft cooperative arrangements with private firms and other research organizations. Three projects especially demonstrate this capability—Lunar Prospector, the X-36 and SOFIA.

Lunar Prospector

Goldin launched NASA's Discovery Program in 1992 to fund highly focused missions with lower costs, shorter timelines, and less risk, by giving the science investigation teams a great deal of freedom. Discovery series projects were meant to reinvigorate the space sciences, which had dwindled as NASA funded Shuttle projects, and to spark public enthusiasm for the continued exploration of space. Discovery mission hardware should be built in less than 36 months, and cost less than $150 million ($250 million including launch costs). Ames' Lunar Prospector was the first competitively selected mission funded under the Discovery Program.

In the 25 years since Apollo, only a few spacecraft have flown by the Moon, and only one had a lengthy encounter. The Clementine spacecraft, built by the U.S. Department of Defense (with scientific management from NASA) to image the lunar surface, orbited the Moon for two months in 1994 in an elliptical orbit no closer than 250 miles to the surface of the Moon. Clementine returned radar signatures that provided indirect evidence of ice crystals at the lunar south pole. Since Apollo era samples showed the lunar regolith to be bone dry, scientists suggested that water was transported to the Moon on comets and asteroids, which created deep craters with permanent shadows that shielded the ice from the Sun's heat.

Spurred by these results, Ames developed plans for a spacecraft to lead NASA's

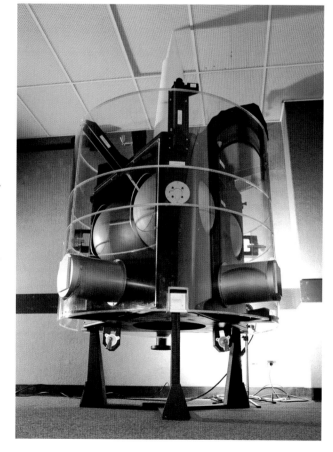

Lunar Prospector prototype in the Ames Space Projects Facility.

Scott Hubbard, NASA mission manager for the Lunar Prospector.

rediscovery of the Moon. Called the Lunar Prospector, it would orbit the Moon for a year, in circular orbit at an altitude of about 60 miles. The idea for the Lunar Prospector was initiated at the Lockheed Martin Missiles & Space Company located adjacent to Ames in Sunnyvale. Former Ames deputy director Gus Guastaferro, then an executive with Lockheed, guided project conception. Ames managed the Prospector contract, and G. Scott Hubbard of the Ames space projects division led all Prospector efforts as the NASA mission manager. The principal investigator was Alan Binder at Lockheed; Tom Daugherty led the team at Lockheed that designed and built the Prospector. (After launch, Binder moved to the Lunar Research Institute of Gilroy, California, to await return of data.) William Feldman of the Los Alamos National Laboratory led the design of three key instruments and the Hewlett-Packard Company built a custom test system using off-the-shelf components. By contracting for parts and services from 25 other Silicon Valley firms, and by

The first operational maps of lunar gravity. Lunar Prospector's Doppler gravity experiment painted these portraits of the near and far sides of the Moon. Peak gravitation is shown in red and valleys in blue.

designing Prospector as a simple spin-stabilized cylinder just 4.6 feet in diameter and 4.1 feet in length, Lockheed took the spacecraft and mission from go-ahead to final test in only 22 months. In addition, Lockheed Martin, at its facility in Colorado, built the Athena launch vehicle which was used for its first time to send Prospector skyward. The total cost to NASA for the mission, including launch, was $63 million. "Prospector has served as a model for new ways of doing business," said Hubbard. "This mission has made history in terms of management style, technical approach, cost management and focused science."[23]

Throughout 1997, Ames built a Prospector mission control room from the operations center that had so long served the Pioneer spacecraft. Mission controllers inserted the Prospector into lunar orbit on 11 January 1998 carrying five science instruments. A gamma ray spectrometer remotely mapped the chemical composition of the lunar surface, measuring concentrations of such elements as uranium, titanium, potassium, iron and oxygen. An alpha particle spectrometer looked for outgassing events that suggested tectonic or volcanic activity. A magnetometer and electron reflectometer probed the lunar magnetic fields for clues about the Moon's core. The doppler gravity experiment, managed by Alex Konoplic of JPL, returned the first lunar gravity map with operational specificity. And a neutron spectroscope, the first ever used in planetary exploration, detected energy flux emanating from the lunar regolith. Hydrogen has a unique neutron signature that is indicative of water ice at higher concentrations. Prospector

Lunar Prospector in orbit around the Moon (artist conception).

returned the first direct measurement of high hydrogen levels at the lunar poles, which Ames scientists believe can only be explained as the presence of water ice.

Ames held a press conference on 5 March 1998 to announce the first science results from Lunar Prospector, only seven weeks after it entered lunar orbit. The indication of water ice embedded in the permanently shadowed craters at the lunar poles made headlines around the world. Future lunar explorers could extract this water for life support or as a source of oxygen and hydrogen fuel. Rough estimates showed up to six billion metric tons of water mixed in fairly low concentrations. After its first year in orbit at sixty miles, Prospector was instructed to swoop down as low as twenty miles to map the Moon at even greater detail. Ames scientists then refined their scientific data and their estimates of water volumes. Mission controllers instructed the Prospector—its fuel now exhausted, its design life far exceeded, and after its 6,800 lunar orbits compiled a complete set of data—to crash into a crater at the lunar South pole on 31 July 1999. Although the impact kicked up no debris visible by ground-based telescopes, NASA scientists using space-based telescopes continued to look for signs of vapor that they could analyze for further evidence of water ice.

NASA/Boeing X-36 Tailless Fighter Agility Research Aircraft

The X-36 proved, with dramatic efficiency, the concept of the tailless fighter. It was conceived in 1989 by researchers at Ames' military technology branch and McDonnell Douglas' Phantom Works in St. Louis (now part of Boeing). It embodied the results of a decade of Ames research into tailless fighters—using wind tunnels, simulators, supercomputers and flight controls. The X-36 lacks vertical and horizontal tails. Instead,

it gets directional stability and flight control through a split aileron and engine thrust vectoring. This innovative design should reduce weight, drag and radar signature and increase the range, maneuverability and survivability of future fighter aircraft.

Rather than build a full-scale piloted prototype, the Ames/Boeing team built a 28 percent scale remotely piloted model. Two X-36 prototypes rolled out in May 1996, only 28 months after go-ahead, at a total project cost of $21 million shared between Ames and Boeing. Each aircraft was 18 feet long, 3 feet high, had a 10-foot wingspan, and weighed 1,250 pounds. They were fully powered by turbofan engines providing 700 pounds of thrust, and flown by a pilot sitting in a ground-station cockpit, complete with a head-up display. By keeping a pilot in the loop, Ames eliminated the expense of complex, autonomous flight controls.

"When we saw this airplane lift off," exclaimed Rod Bailey, the X-36 program manager, "we saw the shape of airplanes to come."[24] Between May and November 1997, the X-36 prototypes flew 31 flights, for over 15 hours, in only 25 weeks. Four different versions of flight control software were tried out. The X-36 reached an altitude of

The remotely piloted X-36 tailless jet fighter test bed in flight over southern California's Mojave Desert.

20,200 feet, and a maximum angle of attack of 40 degrees. The flight tests clearly demonstrated the feasibility of tailless fighters, and showed that they could possess agility far superior to that of today's best fighters.

SOFIA (Stratospheric Observatory for Infrared Astronomy)

SOFIA is the newest generation of airborne infrared observatories—in the tradition of the Kuiper Airborne Observatory, but built from a Boeing 747 and carrying a telescope 2.5 times stronger. Teams of astronomers will be able to observe the radiant heat patterns of space from the cold dark fringes of Earth's atmosphere. At its cruising altitude of 41,000 feet, SOFIA will fly above 99 percent of Earth's obscuring water vapor. Observations impossible for even the largest and highest ground-based telescopes will help answer questions about the birth of stars, the formation of solar systems, the origins of complex molecules in space, the evolution of comets, and the nature of black holes.

Planning for the SOFIA began a decade earlier when the Kuiper was the world's only airborne observatory. Edwin Erickson first nurtured plans to supersede the Kuiper with a bigger and more capable aircraft. Ames space scientists, led by James Murphy, also conceived and developed plans for the liquid helium-cooled Space Infrared Telescope Facility (SIRTF)—the infrared component of NASA's series of great spaceborne observatories. A unique technology group sprang up, led by Craig McCreight and Peter Kittel, to develop low noise

Model of the SOFIA undergoing tests in the 14 foot wind tunnel.

SOFIA during configuration test flights in March 1998.

detectors for SIRTF. When NASA headquarters moved SIRTF management to JPL in 1991, McCreight and Kittel continued their work and Ames revised plans for the SOFIA to complement SIRTF capabilities. Ames aerodynamicists designed and tested ideas for the open air telescope port. During a major upgrade of the information systems for the Kuiper, completed in December 1991, Ames refined the computing and data collection equipment that would be included on the SOFIA. For the next five years, Ames struggled to get funding approved by headquarters and Congress as they reshaped the institutional structure to support SOFIA.

In December 1996, David Morrison, Ames' director of space, announced that Ames had awarded the $480 million SOFIA prime contract to USRA (Universities Space Research Association) of Columbia, Maryland, a private nonprofit corporation with eighty universities as institutional members. USRA was formed in 1970 under the auspices of the National Academy of Sciences to provide a mechanism for collaboration in space exploration. USRA has overall project management, and will later lead scientific operations. The SOFIA contract was a new type of contract—performance based and with full-cost accounting. Unlike previous contracts which specified the resources and personnel a contractor would devote to a project, Ames' contract for the SOFIA specified only the scientific work USRA must accomplish.

"The SOFIA program is a stellar example of NASA's new way of doing business," exclaimed Goldin, "We have taken the parts of a space science program that the private sector can do better and more cost effectively than the government, and had a competitive selection for the privilege of performing those duties."[25]

Modifications to the SOFIA Boeing 747 began in 1998 at Raytheon E-Systems of Waco, Texas, where the aircraft's open cavity was engineered and special equipment installed. Raytheon's lead subcontractor for communication and control software is the scientific systems division of Sterling

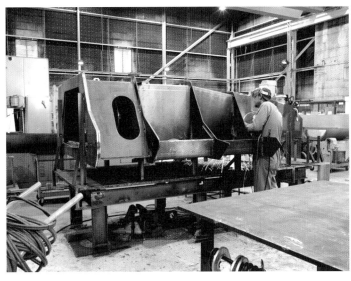

Ames has always had world-class machine shops, where skilled workers can make intricate and reliable sensors or large panels of innovative composite materials. Here an aircraft part is being machined and mounted on a horizontal boring mill.

Software, which brings years of experience in designing and operating computers for the Kuiper. The infrared telescope, 98 inches in diameter, was designed and built by a consortium of Germany's leading aerospace companies—Keyser-Threde GmbH and MAN Technology—managed by the German space agency. Specialized instruments, about 15 per year, will be built by scientists from Ames, the University of California, and other universities. SOFIA's education and outreach program will be conducted in alliance with the Astronomical Society of the Pacific and the SETI Institute. When SOFIA goes operational in 2001, United Airlines will operate and maintain the SOFIA aircraft, which once flew passengers as part of United's fleet. United will train pilots to fly the SOFIA, and will maintain it at its airline hangars in San Francisco or around the world as needed. USRA intends for the SOFIA to make about 160 flights per year of about eight hours each, and base it at Moffett Federal Airfield.

PROCLAIMING THE AMES APPROACH

The success of Ames within the organizational culture of the new NASA was no accident. The Ames way of doing things always involved collaborative research, empowering scientists and engineers, doing things cheaper, getting your hands dirty, and working quickly. As NASA reshaped itself, it looked to Ames and to the leadership style of Harry McDonald as a model of how to do things right. Ames people were called upon to proclaim the reasons for their success, and they looked to their history.

The history of Ames Research Center is reflected in the projects it does and the way it organizes its scientific and technical expertise. The history of Ames is also built into the place. The almost constant whirls and rushes sound out that Ames still operates, as it has for most of its sixty years, the world's greatest collection of wind tunnels. When the update of the Unitary plan tunnels is completed in 2000, all the major Ames simulation facilities will be able to meet the challenges of a new century.

Bill Berry and Rick Serrano, in August 1999, raising the flag that proclaims Ames is ISO 9001 certified for meeting the highest standards of quality and customer satisfaction.

Take a walk around Perimeter Road and see the sun glisten off the big vacuum balls, the vast welcoming sky above the Moffett field runways, and the huge mouth of the 80 by 120 foot wind tunnel ready to gulp air in service of ever-better helicopters. Peek into the high bays, where bustling teams of engineers, programmers, scientists, and human factors specialists build simulators to prove, again, that women and men can always go where they have never gone before if they just think through the trip. Stand at the broad doorways to the shops where skilled machinists craft models of aircraft just conceived, and lace them with incredibly sensitive instruments, while surrounded by racks bulging with the models and parts of the great aircraft and instruments of the past. It's been said that Silicon Valley happened here, because here you could get one of absolutely anything made. That quiet spirit of craftsmanship still thrives around Ames.

See the biologists huddled over sophisticated boxes that will let new generations of tiny ambassadors of earthly flora and fauna grow in space. Follow the glow of screens illuminating the young programmers driving computers ever further and faster. Wander about the library shelves bulging with the reports making even Ames' most abstract theories accessible to all. Glance up at the portraits of Joseph Ames, Smitty DeFrance, and the select members of Ames Hall of Fame, reminding all who enter that, above all, what its history has built into Ames is a respect for all who labor there.

Ames undertook ISO 9001 certification as a chance to align its tradition of excellence with the international standard for quality management. In June 1996, Ames' deputy director William Berry saw how certification benefited work at Great Britain's closest analog to Ames, the Defense Evaluation and Research Agency. When Goldin asked all Centers to fold their Total Quality Management into the broader and better-defined ISO 9001 process, the Ames aeronautical test and simulation division had already begun testing the implementation plan devised for Ames. By 1998, all of Ames embraced the ISO 9001 process as a chance to demonstrate categorically the quality they had so long, and often so quietly, provided to those they

The proposed California Air and Space Center.

served. In April 1999, after an intense review, Ames was ISO certified "without condition," a rare achievement. "When Ames needs to step up we can show superior management process," noted Harry McDonald. "We just don't want too much managerial process."[26]

Ames people started seeing Moffett Field less as a burden, and more as the physical endowment on which to build the Center of their dreams. With leadership from McDonald, Berry and Michael Marlaire, Ames' director of external affairs, Ames people began to view Moffett Field not as a problem to be managed or a collection of historical artifacts from another era of science and technology. Instead, they came to view the Moffett land as a once-in-a-lifetime opportunity—as a large, still-underdeveloped piece of land at the epicenter of the world's most dynamic industrial region. "Our Center's traditional agenda and structure were becoming fundamentally unstable because of the change in the world around us," noted Berry. "Today, no one would build huge wind tunnels here, on land this expensive, and where labor costs are so high. Nor would they surround a major research center with a fence."[27] The San Francisco Bay Area is the most prosperous metropolitan area in the nation. It is the nation's third leading exporter overall, producing more than one-fourth of America's high tech exports. A fifth of the 100 fastest growing global companies are located there—including most of the leaders in computing, communications and biotechnology that collaborate closely with Ames. For Ames to continue to flourish, to advance knowledge, and to contribute to the national well-being, the Center's leadership realized it must be firmly rooted in that community. They explored ways to use this endowment of land to bolster that connection.

Throughout 1998 Ames hashed out the details of a bold new development plan. The Ames portion of the base will remain fenced and operate as before. The airfield will remain intact though quiet. Then in the old Navy portion of the base there will emerge a new complex of research buildings. Stanford University needs research facilities and the University of California at Santa Cruz needs space for extension education. So these universities will help develop the land—perhaps into something akin to the Stanford Research Park—while Ames will control the improvements. The universities will also bring in industrial partners and, perhaps in the coming decades, take a larger role in managing the research done at Ames. "Reimbursable" Space Act tenants—mostly start-up companies helping to transfer NASA technology, will pay Ames for supporting the

NASA Ames Research Center senior staff in January 1998: (standing, left to right) Kenneth M. Ford, associate director for information technology; Jana M. Coleman, director of Center operations; Robert Rosen, associate director for aerospace programs; Lewis S.G. Braxton III, chief financial officer; Robert J. Hansen, deputy director for research; David Morrison, director of space; Nancy J. Bingham, assistant director for strategic planning; (seated, left to right) Steven F. Zornetzer, director of information systems; Harry McDonald, director; William E. Berry, deputy director; John W. Boyd, executive assistant to the director.

At the September 1997 Ames Community Day and Open House, Ken Souza of the Ames life sciences division explains the need for spacesuits like those designed at Ames. More than 220,000 people—young and old—stood fascinated by the technology displays.

infrastructure of the complex. The focal point will be the California Air and Space Center, a science education facility the size of Hangar One. Five acres were set aside for The Computer History Museum. The south gate will be reopened to allow easy, uncontrolled public access to an open campus.

Ames held its first open house community day on 20 September 1997. Thousands were expected; nearly a quarter million of Ames' closest friends streamed in. Ames displayed its latest technology at 17 sites around the Center, including demonstrations of the Mars rover and many of its wind tunnels. "Partnership" unified the 150 exhibits

inside Hangar One, where local schools, companies, federal agencies, and community organizations bragged about all they had accomplished by working with Ames. Over 1,300 Ames ambassadors helped the crowd, describing the science behind the dazzling displays. "We all witnessed actions so extraordinary," effused Lynn Harper, who coordinated the space sciences exhibits, "that we thought we'd burst with pride."[28] As David Morse and Donald James, the Ames external affairs co-chairs who so quickly organized the open house, walked around to check on things, people applauded.

Morale at Ames had sunk low in recent years—budget cuts by Congress, the transfer of programs to other Centers, neglect and scolding from headquarters, and a lack of technical leadership within Ames. Most Ames people thought they had nothing to show the public but relics of its past. As they caught glimpses of the public interest in the open house, however, enthusiasm grew. Two weeks before the event it seemed like half the people at Ames were working on it. With the extraordinary turnout, employee morale skyrocketed and has risen steadily since. The open house displays and demonstrations let Ames shed the trappings of its past and embrace its future by declaring—loudly, visibly and harmoniously—how it was stepping up to its mission responsibilities in information technology, astrobiology and aviation capacity and safety. This time Dan Goldin, who had inspired the event after he met with local leaders six months before, had to compose himself as he welcomed the throngs so fervently interested in all Ames had contributed to its community. Ames director Harry McDonald reflected:

"September 20, 1997 was truly a momentous day in the life of Ames Research Center—a day when we made history and recast the course of our future. Together, as we transform this incredible Center, we are reinventing ourselves in the process. The sense of excitement is obvious and evident everywhere. Our workforce has a new sense of pride. A better, more robust Ames will be our legacy; effecting the transformation is our reward. Community Day did not initiate this process. But, as we look back, it will always stand as the most visible signpost on the historic pathway of change, and the point from which all future progress will be measured. Collectively, we have changed both the perception and reality of Ames."[29]

1 8 6 4   1 9 4 3

JOSEPH SWEETMAN AMES

Appendix

To commemorate the 25th anniversary of its first meeting, the NACA convened on 18 March 1940 in Washington, D.C. The meeting was largely ceremonial, to reflect on how well they had built their organization. "You are doing a great job," effused President Franklin Roosevelt. "You are one governmental agency that doesn't give me a headache."[1] The Committee transacted only one item of business. Two, perhaps, if you include naming a small committee to carry this notice to Joseph Sweetman Ames: "The aeronautical research laboratory, authorized by Act of Congress on August 9, 1939, and which is now under construction at Moffett Field, California, will be named in your honor and will be known as the 'Ames Aeronautical Laboratory.'"[2]

Illness forced Ames to retire from the NACA months before, in October 1939, after 24 years of service. The NACA had become the world's greatest institution for aeronautical science largely because of Ames' dignified leadership and his devotion to basic research—two traits he had cultivated at the other great institution he helped build—The Johns Hopkins University.[3]

As to the trustees it is their absolute duty to accept any recommendation
that comes to them from the faculty when proposed by the president.
As for the president, his primary duty is to uphold the faculty.
I refer to tenure of office, freedom of speech, morale,
all that goes to make up the faculty.... The primary purpose in the life of
a professor is to conduct his own investigations and lead his own scholarly life,
and the more attention he can pay to that,
the better it is for an institution.

J. S. Ames, Address before the American
Association of University Professors, December 1929

AMES AT JOHNS HOPKINS UNIVERSITY

Joseph Sweetman Ames was born in Manchester, Vermont, on 3 July 1864, into a long line of Yankee industrialists and educators. His uncle, from whom he took his middle name, was president of Union College. His father, a physician, moved Ames and his mother to Niles, Michigan, but died in 1869. Ames' mother, a strong woman who was devoted to literature, then moved him to Faribault, Minnesota, to be part of the cultural community built amid those rolling hills by the Protestant Episcopal Bishop Henry Benjamin Whipple. Ames was a star pupil at the Shattuck School, where he got the classical, humanist education that made his speaking and writing so clear, logical and persuasive.

Ames recalled happening upon, at the age of 15, an article in a popular magazine that described the newly founded Johns Hopkins University: "It possesses no history, claims no distinguished sons, has indeed hardly reached the dignity of alma mater," but it was dedicated to "raising the level of educational standards" through "original research."

Ames arrived at Johns Hopkins as a freshman in 1883, and died as its president emeritus sixty years later. He spent only one year away: as a post-doctoral student visiting the great laboratories of Europe and attending the physics lectures of Hermann von Helmholtz in Berlin. In fact, Hopkins was modeled on the German ideal of education through research, seminars, and the progression from undergraduate to graduate studies. Hopkins was only three years old and had already stirred controversy among American educators when Ames arrived.

Ames gravitated toward the laboratory of Henry Rowland, a pioneering spectroscopist with a knack for professional organization. Rowland named Ames his assistant in 1888, Ames took his Ph.D. in 1890, then quickly assumed greater duties teaching physics and managing the laboratory. Ames was most effective at getting good work from his students. On Rowland's death in 1901, Ames officially became laboratory director. As a researcher, Ames developed spectroscopy, a dramatic new tool for analyzing the composition of materials and, later, the structure of the atom. Ames attended the birth of astrophysics when he was invited by the astronomical director of the Naval Observatory to supervise the spectroscopic work for a 1900 eclipse expedition.

Ames took teaching and administration seriously, and his own experimental program suffered. Yet he avidly followed the literature (he always called himself a student of physics rather than a physicist) and thus always had insights with which to help his students overcome their experimental roadblocks. As his laboratory expanded in scope, so did his teaching. In 1915 he crafted a three-year course (published as the textbook *General Physics*) that laid out the architecture of physics in a period of rapid change. Ames was among the first American physicists to risk their reputations by publicly defending Einstein's theory of relativity. Well into old age, Ames was perceived as one of the "younger generation" of physicists because of his willingness to support fresh ideas. One young man Ames mentored was Hugh Dryden, who in 1949 became the first person to fill the new, full-time post of director of the NACA (where he remained until the NACA was absorbed into NASA in 1958). Ames' lecture style was formal, complete, logical and clear. (His personal conversation likewise was kind-hearted and fair but business-like and blunt—the result, some speculate, of a lifelong effort to control a stammer.) Ames only stopped teaching in 1925 when named dean. He was named provost in 1926.

Ames was 65 years old when he was selected president of Hopkins in 1929; he knew it would be a trying time. The university was moving into its new campus, built for $4.2 million in the Homewood section of Baltimore, just as the Great Depression wiped out Hopkins' endowment. Ames saved the university from bankruptcy by valiant appeals to the alumni. Ames also inherited a fractious faculty debate over the Goodnow plan, which threatened to sever the traditional link at Hopkins between graduate research and undergraduate teaching. And Ames dealt with increasing disciplinary problems by creating a structure for students to discipline themselves. "No administrative officer could be more accessible and none could take greater pains to gain a sympathetic understanding of the problems of others," wrote a journalist on Ames' presidency. "These are valuable qualities in the administrator of so loose a confederacy as a university."[4]

When his wife died in 1931, Ames committed himself to work and community service. (Though he had always tried to "keep my fingers on the city's pulse without the patient knowing it.") He was elected a lay member of the standing committee of the Protestant Episcopal Diocese of Baltimore, and continued to serve as president of the Baltimore Country Club. Since 1900 Ames had directed a series of public lectures in physics for Baltimore area teachers. He accepted appointment to the Baltimore school board in 1932, then led opposition to a 1935 state bill requiring anyone teaching in Maryland to take a loyalty oath. His reasoning was pragmatic. Hopkins hosted many eminent foreign scholars, and the bill "ignores the fact that scholarship is a world-wide affair." But Ames attacked the bill as a larger evil: "Patriotism need not be taught by law." The press called him a man of courage, veterans supported him, as did those who championed Maryland's traditional tolerance of beliefs. "I would select as the fundamental principle of Americanism," Ames said, "freedom of thought, freedom to express one's best conclusion as to what is truth." Maryland's governor vetoed the loyalty oath bill.

In his final commencement as president of Hopkins, Ames told his students: "My hope is that you have learned or are learning a love of freedom of thought and are convinced that life is worth while only in such an atmosphere....It is doubtful if you ever attain absolute truth, and it is certain that you will never know if you do attain it. This should, however, be your constant endeavor."[5]

Freedom of inquiry, and the responsibility that it entails, are values Ames also instilled into the National Advisory Committee for Aeronautics (NACA).

AMES IN THE NACA

Ames was a founding member of the NACA, appointed by Woodrow Wilson in 1915. His research management skills and zeal for public service probably led to his appointment. So did his proximity to Washington: all NACA members worked part-time, without compensation. Mostly, Ames was appointed to represent physics. As aeronautics struggled to become a science, many disciplines claimed aeronautics as theirs. Looking at the problems of predicting how a solid body moves through the atmosphere, Ames wrote, "These matters make up the subject of aeronautics; and in order to investigate them the same methods must be applied as in any department of physics....When the physical facts are known, the engineer can design his aircraft, the constructor can make it, and the trained man can fly it; but the foundation stone is the store of knowledge obtained by the scientist."[7]

Ames leapt to take on the NACA's most challenging assignments. He chaired the Foreign Service Committee of the newly founded National Research Council and, in 1917, toured Britain and France to study the organization of science in service of the war. Ames oversaw the NACA's patent cross-licensing plan that allowed manufacturers to share technologies, and

> "When you were first appointed by President Wilson in 1915, very little was known about the science of aeronautics. To you and your colleagues were entrusted by law the supervision and direction of the problems of flight....The remarkable progress for many years in the improvement of the performance, efficiency, and safety of American aircraft, both military and commercial, has been due largely to your own inspiring leadership in the development of new research facilities and in the orderly prosecution of comprehensive research programs."
>
> F. D. Roosevelt to J. S. Ames,
> 10 October 1939[6]

his unquestioned impartiality and integrity made the plan work. Ames expected the NACA to encourage engineering education. Older men sat on the NACA's nested hierarchy of committees, mostly to facilitate the work and careers of younger men. Ames pressed universities to train more aerodynamicists, then structured the NACA to give young engineers on-the-job training.

Ames gave the NACA its scientific "tunnel vision" then valiantly protected that portion of the burgeoning aircraft effort. Ames led a special committee on the NACA's postwar organization, fashioned a small but effective NACA central bureau, and urged the Committee to voice a vision and program for research. The NACA focused on research rather than policy. The military services and, later, the Commerce Department decided what aircraft to buy and how to use them; the NACA helped make better aircraft. And Ames decided that NACA research mattered most in aerodynamics. He championed the work of theorists like Max Munk. The extraordinary wind tunnels at Langley Aeronautical Laboratory reflected Ames' vision, as well as the faith Congress placed in him. The extraordinary work done there resulted in part from Ames' frequent visits to assess and encourage the work of its young staff.

A quiet triumvirate made the NACA so effective. John Victory administered its offices and committee hierarchy. Engineer George Lewis, NACA director of aeronautical research, connected NACA work with the needs of the military and industry. And Ames travelled to Washington one day each week, kept an eye on politics, and made it known that the NACA was about science. "The quiet, conservative, methodical style of the Committee," wrote historian Alex Roland, "can be attributed in large measure to this gentle man."[8]

Ames became chair of the NACA main committee in 1927. Two years later, he accepted the Collier Trophy on behalf of the NACA. He kept the NACA alive when Herbert Hoover tried to kill it and transfer its duties to industry. And Ames urged funding for a second laboratory.

Ames accepted a nomination, by Air Minister Hermann Göring, to the Deutsche Akademie der Luftfartforschung. Ames then considered it an honor—many Americans did—and was surprised to learn about the Nazis' massive investment in aeronautical infrastructure, then six times larger than the NACA. Ames more fervently advocated an expansion of NACA facilities to prepare for war.

A stroke in May 1936 paralyzed the right side of his body so, at age 73, he retired to the garden office at his home on Charlecote Place in Guilford near Baltimore. He immediately resigned as chairman of the NACA executive committee and, in October 1937, resigned from the NACA main committee. But Ames grew concerned when Congress bogged down the NACA proposal for its Sunnyvale laboratory. In a well-reasoned letter to Clifton A. Woodrum of the appropriations committee, Ames asked why America should spend hundreds of millions of dollars on second-best aircraft. "What makes the project emergency in character is the fact that Germany, because of her larger research organization, now has the ability to design, and actually has in service, aircraft of superior performance." Ames concluded with a rare, personal appeal: *"For nearly twenty years, I have been appearing before the Appropriations Committee. I have supported what at times appeared to be bold plans for development of research facilities in the United*

Professor William F. Durand stands at the center of an impressive group of guests, on 8 June 1944, for the dedication of the NACA Ames Aeronautical Laboratory.

Ames chairing the April 1929 meeting of the NACA Main Committee. (Left to right): John F. Victory, William F. Durand, Orville Wright, George K. Burgess, Brigadier General William Gillmore, Major General James Fechet, Joseph S. Ames, Rear Admiral David W. Taylor, Captain Emory Land, Rear Admiral William A. Moffett, Samuel W. Stratton, George W. Lewis and Charles F. Marvin.

States. I have never supported a padded or extravagant estimate. I have never supported a project that the Congress refused to approve. And every such project has proved successful for its purpose. Now, through impairment of health I am nearing the end of my active career. I have served as a member of the National Advisory Committee on Aeronautics for nearly twenty-five years, without compensation. I ask nothing from my country in the way of reward. My compensation has been tremendous in the satisfaction that has come to me from the realization that the work of the National Advisory Committee for Aeronautics has been successful over a long period of years in enabling American manufacturers and designers to develop aircraft, military and commercial, superior in performance, efficiency, and safety to those produced by any other nation. Now, I regret to say, the picture is changed. I still, however, have faith in our ability, with your support and the support of Congress, to regain for America the leadership in scientific knowledge which will enable our designers and manufacturers again to produce superior aircraft."[9]

On 8 June 1944, once the administration building was complete, the NACA officially dedicated its new laboratory in Sunnyvale to Joseph Sweetman Ames. Ames had died a year before. He never set foot in the laboratory that bears his name. In a letter to Stanford professor William Durand, who hosted the dedication ceremony, General Henry H. "Hap" Arnold called "Dr. Ames the great architect of aeronautical science…It is most appropriate that it should now be named the Ames Aeronautical Laboratory, for in this Laboratory, as in the hearts of airmen and aeronautical scientists, the memory of Joseph S. Ames will be enshrined as long as men shall fly."[10]

Acknowledgments

Jack Boyd has played a unique role in the history of Ames, much like the role he played in producing this history of Ames. Jack joined Ames in 1947 as a specialist in supersonic aerodynamics. In 1966 he became technical assistant to Ames' director and since then—as he moved up to deputy director, associate director, NASA associate administrator for management, and well into his so-called retirement years—Jack has served as Ames' institutional memory and the embodiment of its ethos of personal respect. When recruiting leaders from outside of Ames, like Hans Mark and Harry McDonald, NASA headquarters asked Jack to lead the recruits on their first tours of the place. When Ames people struggling to build their programs needed insight into the cultures and personalities around NASA, they could rely on Jack's diplomacy and his friendships throughout the agency. He found funding for this history, introduced me to people of every generation around Ames, and helped me interpret both the bright and dark years of Ames history. Jack has always devoted time to education and outreach. Most important, Jack lives the belief that a strong sense of the past is great grounding in the rush to forge the future.

David Morse, the person at Ames responsible for communicating knowledge, oversaw this history and taught me much about how to write about Ames. Laura Shawnee efficiently managed the contract for Ames; Annette Rodrigues, Doreen Cohen and Sheila Keegan managed the contract for Quantum Services. Carla Snow-Broadway helped find documents; Lynn Albaugh helped find photographs; Jeanette Louis-Sannes helped with office stuff; and Dan Pappas, the Ames history liaison, helped find library materials. Mary Walsh supervised the publication team, Charlotte Barton sweated the review process, Boomerang Design Group did the graphic design and John Adams managed the printing.

Walter Vincenti gave great advice, not only as a world-renowned historian of technology at Stanford University, but as an aerodynamicist and tunnel builder during Ames' first eighteen years. John V. Foster and John Dusterberry welcomed me into the Owl Feathers Society. Carl Honaker and Mike Makinnen told me much about the Navy side of Moffett Field. Stephanie Langhoff, Ames chief scientist and leader of the Ames Hall of Fame project, taught me how working scientists view Ames' history.

Helen Rutt served as project archivist, collected and processed a reference collection on Ames history, and compiled finding aids to the primary materials. Kathleen O'Connor takes superb care of those documents stored at the National Archives and Records Administration, Pacific Sierra Region in San Bruno. Karen Dunn-Haley wrote on human resources, safety, and environmental issues at Ames and compiled the bibliography. Kristen Edwards wrote on the history of Ames/Soviet collaboration on the Cosmos/Bion program. Mark Wolverton researched the role of university principal investigators in Pioneers 10 and 11.

More than seventy past and present Ames employees shared their time and thoughts with me, both formally and informally. Many of these oral history interviews are transcribed and deposited in the Ames history collection, though I have only cited those quoted directly. Thanks especially to Scott Hubbard, Nancy Bingham, Eugene Miya, Alvin Seiff, Howard Goldstein, Warren Hall, Michael McGreevy, Jeffrey Cuzzi, and Bill Berry.

This is a work for hire, though the only restrictions Ames placed on this manuscript was that it be done in time to celebrate their sixtieth anniversary. Some will argue that I failed to give adequate attention to significant projects, failed to give sufficient credit to everyone who dedicated their lives to the institution, or failed to capture all the struggles they had to overcome. A great many people have reviewed this manuscript for errors of fact, interpretation and balance, but any that remain are mine alone.

Ames' director Harry McDonald was a superb audience because he conveyed his intense curiosity—as he does so well with almost everybody's work—of learning more about Ames' past. My thanks to everyone who said they looked forward to reading what I wrote. Most made sure that I understood, however, that the best scientists, engineers and managers will keep coming to Ames—not because of its history of success that I was tasked to portray—but to work with the best minds and the best tools in their professions.

Bibliographical Essay

Two histories of Ames precede mine, and both were valuable sources in writing this history. The chapter on Ames as a NACA laboratory is based largely on Edwin P. Hartman, *Adventures in Research: A History of Ames Research Center, 1940-1965* (NASA SP-4302, 1970). Hartman directed the NACA field office in Los Angeles from 1940 to 1960, meaning he led Ames outreach when the audience that most concerned Ames was engineers in the aircraft industry. I also relied upon a series of memoranda summarizing Ames contributions, written by Ames branch chiefs, compiled by Manley J. Hood in February 1960, at the request of John F. Victory, and filed with the history collection in the vault of the Ames main library. On deicing work see Glenn E. Bugos, "Lew Rodert, Epistemological Liaison and Thermal De-Icing at Ames," in Pamela Mack, ed. *From Engineering Science to Big Science: The NACA and NASA Collier Trophy Research Project Winners* (NASA SP-4219, 1998) 29-58; on the blunt body concept see H. Julian Allen and A. J. Eggers, Jr., "A Study of the Motion and Aerodynamic Heating of Ballistic Missiles Entering the Earth's Atmosphere at High Supersonic Speed" (NACA TR 1381, 1958); on wind tunnels around the NACA see Donald D. Baals and William R. Corliss, *The Wind Tunnels of NASA* (NASA SP-440, 1981). On formation of the second laboratory within the NACA, see Alex Roland, *Model Research: The National Advisory Committee for Aeronautics, 1915-1958* (NASA SP-4103, 1985).

The chapter on Ames' transition into NASA relies again on Hartman's history, as well as on Elizabeth A. Muenger, *Searching the Horizon: A History of Ames Research Center, 1940-1976* (NASA SP-4304, 1985). In addition, in February 1976, Edith Watson Kuhr compiled a series of historical memoranda written by Ames branch chiefs, and kept in the Ames history collection. On the introduction of the life sciences see John Pitts, *The Human Factor: Biomedicine in the Manned Space Program to 1980* (NASA SP-4213, 1985). The best histories of the Pioneers are Richard O. Fimmel, James A. Van Allen, and Eric Burgess, *Pioneer: First to Jupiter, Saturn, and Beyond* (NASA SP-446, 1980); Richard O. Fimmel, William Swindell, and Eric Burgess, *Pioneer Odyssey* (NASA SP-349, 1977); William E. Burroughs, *Exploring Space: Voyages in the Solar System and Beyond* (Random House, 1990); and William R. Corliss, *The Interplanetary Pioneers* (NASA SP-278, 279 and 280, 1973).

The chapters on Ames since the 1970s are based largely upon materials found in the history collection at the Ames main library. *The Ames Astrogram* is the Ames employee newsletter and the best source on everything happening at Ames. The collected press releases issued by the Ames external affairs office do a superb job of explaining media-intense activities like space probe encounters.

Since Ames researchers appreciate that they are making history, they have written a good many histories of their work. Most of these are for technical audiences and address specific projects. Ames' aircraft and rotorcraft projects are nicely summarized in Paul F. Borchers, James A. Franklin, and Jay W. Fletcher, *Flight Research at Ames: Fifty-Seven Years of Development and Validation of Aeronautical Technology* (NASA SP-1998-3300). On airborne astronomy see Wendy Whiting Dolci, "Milestones in Airborne Astronomy: From the 1920s to the Present," *AIAA Reprint 97-5609* (American Institute of Aeronautics and Astronautics, 1997); on computational fluid dynamics see Ames Research Center, *Numerical Aerodynamics Simulation* (NASA EP-262, 1989). The best history of Ames' contributions to VTOL aircraft is Martin D. Maisel, Demo J. Giulianetti, and Daniel C. Dugan, *The History of the Tilt Rotor Research Aircraft, from Concept to Flight* (NASA Monographs in Aerospace History, No. 17, 1999). See also David D. Few, *A Perspective on 15 Years of Proof-of-Concept Aircraft Development and Flight Research at Ames-Moffett by the Rotorcraft and Powered-Lift Flight Projects Division, 1970-1985* (NASA Reference Publication 1187, 1987); G. Warren Hall, *Flight Research at NASA Ames Research Center: A Test Pilot's Perspective* (NASA TM-100025, 1987), and Hans Mark, "Straight Up Into the Blue," *Scientific American* 277 (October 1997) 78-83.

In addition, many boxes of primary materials are stored at the Pacific Sierra regional facilities of the National Archives and Records Administration in San Bruno, California. The records of Ames during the NACA years are well organized and indexed. The records from 1958 to 1976 have been transferred to the National Archives, though they are not well indexed. Virtually all records since 1976 remain with the Federal Records Center or on-site at the Center. The available indexes can be found at the website for the California Digital Library.

A more complete guide to all materials available for researching topics in Ames history can be found at the website for the NASA Ames history project at http://history.arc.nasa.gov. This includes a research bibliography, list of Ames award winners, guides to primary materials at the National Archives and in the Ames main library, guides to materials at NASA headquarters, and list of interviewees.

ENDNOTES

These endnotes provide citations only for direct quotations. For sources for further reading, see the bibliographic essay, or the Ames history web site at http://history.arc.nasa.gov.

CHAPTER 1

[1] Alex Roland, *Model Research: The National Advisory Committee for Aeronautics, 1915-1958*, vol. 1 (NASA SP-4103, 1984) 154.

[2] Roland, *ibid.*, 160.

[3] Interview with Walter G. Vincenti, 16 June 1999.

[4] "In Memoriam: H. Julian Allen, 1910-1977," *Astrogram* (10 February 1977) 1-3.

[5] *Op cit*, note 3.

CHAPTER 2

[1] Interview with Clarence Syvertson, 7 August 1999.

[2] Hans Mark, "Pioneer Jupiter-Saturn: Better, Cheaper, Faster in 1970," in *12th National Space Symposium Proceedings Report* (United States Space Foundation, 1996) 63-67.

CHAPTER 3

[1] "NASA Focusing on June Date for Merger Recommendations," *Aviation Week and Space Technology* (1 June 1981) 23.

[2] Kevin G. Tucker, "Aviation Reporting System Saves Lives," *Astrogram* (7 June 1991) 2.

[3] Jane Hutchison, "Unique Ames Facility Aids Space Research," *Astrogram* (28 February 1992) 2.

[4] Hans Mark and Robert R. Lynn, "Aircraft Without Airports: Changing the Way Men Fly," *Vertiflite* 34 (May/June 1988) 80-87; Robert R. Lynn, "The Rebirth of the Tilt Rotor: The 1992 Alexander A. Nikolsky Lecture," *Journal of the American Helicopter Society* (January 1993) 3-14.

[5] Michael Mewhinney, "Final Chapter Closes on History Making QSRA Project," *Astrogram* (18 March 1994) 4.

[6] Donald James, "RASCAL Off and Flying," *Astrogram* (27 March 1992) 1.

[7] Correspondence with James O. Arnold, 20 September 1999.

[8] Breck Henderson, "Ames Researchers Earn Largest Space Act Award Ever," *Astrogram* (23 July 1993) 2.

[9] "Ames Honored for Space Station Habitability Research," *Astrogram* (28 July 1989) 1.

[10] "Ballhaus to Resign as Director of NASA's Ames Research Center," Ames Press Release 89-54 (6 July 1989).

[11] Ted Shelsby, "Taking the Helm at Martin," *The Baltimore Sun* (3 May 1993) 12c.

[12] William F. Ballhaus, "A message from the Director's office," *Astrogram* (14 July 1989) 2.

CHAPTER 4

[1] William J. Broad, "TRW Executive Chosen to Lead Space Agency in New Directions," *New York Times* (12 March 1992).

[2] "NASA Administrator Daniel S. Goldin Presents His First Employee Address," *Astrogram* (8 May 1992) 3-6.

[3] Interview with Warren Hall, 12 July 1999.

[4] Breck W. Henderson, "NASA Headquarters Conducts Security Raid at Ames Center," *Aviation Week and Space Technology* (17 August 1992) 26.

[5] *Op cit*. note 2.

[6] Ann Hutchison, "Goldin Praises Research at Ames During His Recent Visit," *Astrogram* (28 October 1994) 1+.

[7] Bob Rosen, "Quality Improvement Plan Approved by Center Board of Directors—Implementation to Begin Shortly," *Astrogram* (2 April 1993) 6.

[8] Palmer Dyal, "Employee Survey, Malcolm Baldridge Self-Assessment Completed," *Astrogram* (16 September 1994) 2; *Total Quality Survey: NASA Ames Research Center* (Mountain States Solutions and Novations Group, Inc., January 1994).

[9] "Appendix 1—Cultural Climate and Practices Review Team Report, July 27-31, 1992," in *Cultural Climate and Practices Plan: NASA Ames Research Center* (March 1994) 61.

[10] Melody Petersen, "Ames Ends Tough Year with 2 Key Retirements," *San Jose Mercury News* (11 December 1993).

[11] Cover Letter, *Cultural Climate and Practices Plan: NASA Ames Research Center* (March 1994).

[12] Interview with Nancy Bingham, 15 July 1999.

[13] David Perlman, "NASA Proposal Would Shift Ames' Work to Colleges, Firms," *San Francisco Chronicle* (4 May 1995) A1.

[14] Kenneth J. Szalai, "Dryden's Director Sends Appreciation Message to Ames," *Astrogram* (1 April 1994) 1.

[15] "Review Team Proposes Sweeping Management, Organizational Changes at NASA" (NASA Headquarters Press Release No. 95-73, 19 May 1995).

[16] "Press Release: Eshoo Calls for GAO Investigation of NASA Plan to Move Aircraft from NASA/Ames and Other Facilities" (Office of Congresswoman Anna Eshoo, 26 September 1995).

[17] Interview with Harry McDonald, 9 December 1998.

[18] Michael Mewhinney, "Telepresence Technology to Study Antarctic Lakes," *Astrogram* (9 October 1992) 4.

[19] Donald James, "State of the Art Simulator To Be Installed in MVSRF," *Astrogram* (22 May 1992) 1+.

[20] Donald James, "Ames Technology Could Improve Air Traffic Control," *Astrogram* (14 August 1992) 1+.

[21] Interview with Stanton Harke, 28 January 1999.

[22] David Morse, "Goldin Lays Out Vision for NASA Astrobiology," *Astrogram* (24 May 1999) 1+.

[23] "Lunar Prospector Program Profile," *Lockheed Martin Missiles & Space Press Release* #98-41 (March 1998).

[24] Michael Mewhinney, "Remotely Piloted Tailless Aircraft Completes First Flight," *Astrogram* (30 May 1997) 10.

[25] David Morse, "SOFIA 'Privatization' Contractor Team Selected," *Astrogram* (10 January 1997) 1.

[26] Henry McDonald, "State of the Center Address," 7 July 1999.

[27] Interview with William E. Berry, 12 February 1999.

[28] David Morse, "Community Response Is 'Off The Scale,'" *Astrogram* (3 October 1997) 3.

[29] Henry McDonald, "A Message From The Director," *Astrogram* (3 October 1997) 1.

APPENDIX

[1] J.F. Victory to J.S. Ames, 25 April 1940 (File 3: Miscellany: Box 1: G5002: Maryland Historical Society, Baltimore, Maryland).

[2] Edward Warner, Chairman, Committee on Notification to Joseph S. Ames, 17 April 1940 (File 5: Box 8: G5002: Maryland Historical Society).

[3] The best biography of Ames is by Henry Crew, "Biographical Memoir of Joseph S. Ames," *Biographical Memoirs: National Academy of Sciences* XXIII (1944) 181-201. A former student wrote a biography detailing Ames' faults as a person, teacher and physicist; N. Ernest Dorsey, "Joseph Sweetman Ames, The Man," *American Journal of Physics* 12 (June 1944) 135-148; N. Ernest Dorsey, "Joseph S. Ames and Aeronautics," *Johns Hopkins University Alumni Magazine* XXXII (June 1944) 93-44. See also Lyman J. Briggs, "Obituary: Joseph Sweetman Ames," *Science* (30 July 1943) 100-101; "Dr. J.S. Ames Dies, Led Johns Hopkins," *New York Times* (25 June 1943) 17.

[4] P. Stewart Macaulay, "The President of Hopkins Retires," The Baltimore *Sun Magazine* (2 June 1935) (in vertical file number 4, Special Collections, Milton S. Eisenhower Library, The Johns Hopkins University, Baltimore, Maryland).

[5] J.S. Ames, "Commencement Address: June 11, 1935," *Johns Hopkins University Alumni Magazine* 24 (June 1935) 1-6.

[6] Franklin D. Roosevelt to Joseph S. Ames, 10 October 1939 (File 5: Box 8: G5002: Maryland Historical Society).

[7] Joseph S. Ames, "Aeronautic Research," *Journal of the Franklin Institute* (January 1922) 15-28.

[8] Alex Roland, *Model Research: The National Advisory Committee for Aeronautics, 1915-1958* (NASA SP-4103, 1985) 169.

[9] J.S. Ames to Clifton A. Woodrum, 27 March 1939 (Folder 1: Box 1: Record Group 255.E13: Archives II: National Archives and Records Administration).

[10] Letter of Henry H. "Hap" Arnold to William F. Durand, 26 May 1944 (Folder 1: Box 1: Record Group 255.E13: Archives II: National Archives and Records Administration).

Photo Index

Front Matter
vi: ACD97-0225
vii: AC99-0126-1
viii: AC98-0172

Introduction
1: A-13425
2: AAL-5528, AAL-1794, AAL-1800
3: AAL-3703, AAL-3803

Chapter 1
4: A-8051
5: AAL-4960
6: A91-0261-18
7: AC71-2761
8: A93-0073-3, William Moffett (courtesy Moffett Field Historical Society)
9: 95-HC-379 (*Fluid Dynamics*, by Tina York), M-831
10: A-14870, A93-0074-8
11: A-35560, G-325-32-0-82
12: G-616, G-492, M-253
13: G-597, AAL-3782
14: A93-0073-7, AAL-1166, AAL-5010A
15: C47_20298, A-9896
16: A-12679, M-921
17: AFST-37, AAL-1449
18: A-13516, A-8087
19: A-6538, AAL-2265, A-8644
20: A-10060, A-22147
21: A-9943
22: M-704-2, A-22882
23: AAL-5985
24: M-704-1
25: A-8479
26: A-22548
27: A-9364, A-17857-14
28: A-21303, AC-28536
29: A-11807, A-8822
30: A-22817
31: A-23928, A-28249
32: A-10071, A-13703, A-12995
33: A-22437
34: RT Jones, A-13146-8
35: A-13011
36: A-35036

37: A-22498
38: A-18830, A-20416.1
39: A-16545
40: A-20393.1

41: A-24604, A-8ft-SSWT-355, A-8ft-SWTT-356
42: A-14376
43: A-14294, A-15481
44: A-15058.1
45: A-22970
46: A-15590, A-15768
47: A-17181-B
48: A-23703
49: EC68-1889

Chapter 2

50: A-22664
51: AC-42137
52: A-36042-2.1
53: 69-HC-895, AS11-40-5878
54: AC-29418-6, A-41727-6-4
55: A-31071
56: AC-42117-40, A-34009
57: A-24953, A-35501-17
58: A-23730, A-23753
59: A-29007
60: A-23438, A-28418
61: A-24013
62: AC78-1071
63: A-41445-1, A-37250
64: A-28917-2
65: A-36545
66: AC-31392, A-33038-22, A-26921-B
67: A-28256, A-28326-36
68: A-33996, A-37700

69: A-23856.1, A-29323
70: AC-28968, AC-31542
71: AC79-0198-15
72: A-36611, A-29724-2
73: A-36675-125, AC-40699
74: AC-30133, AC71-6490
75: 87-HC-151 (*Tracking*, by Hans Cremers)
76: AC-42375-3, AC77-0893-5
77: A-33753-2
78: AC88-0741 (*Evolution of Life*, by Robert Bausch)
79: 69-HC-1353, AS12-47-6921
80: A-31214
81: AR-1632B-452
82: A-35069
83: EC66-1567, A-23796
84: A-29258
85: AC-30351-2, AC99-0222-1
86: A-33228 (*Pioneer 6*, by T Howard), A-33198
87: Pioneer 10/11.2
88: AC72-2139, AC74-9032-236
89: A71-8747
90: AC72-1338, AC71-8744
91: 88-HC-37 (*Jupiter Flyby*, by Martin Berkon), A73-9044-2
92: AC73-9175, AC97-0036-3 (Art by Thomas Esposito), AC97-0036-1 (Art by Thomas Esposito)
93: AC83-0378-40, AC73-9034 (Art by Chris Chizanskos)

94: AC74-9009, AC74-9006 (Art by Rick Guidice), 80-HC-251
96: AC79-9180, A73-4898
97: Pioneer 10th Pin (Art by Cheryse Triano)

Chapter 3
98: AC91-0667-2
99: AC92-0326-24
100: AC99-0222-2
101: AC86-0410-3
102: AC75-0450, AC84-0099-21
103: A76-0667-11
104: AC74-2201, A75-0962-7
105: AC97-0211-3
106: AC70-1343, AC99-0222-3
107: AC96-0231-1
108: EC97-44165-35
109: AC84-0081, AC75-0261-1
110: AC90-0652-2
111: AC92-0437-3
112: AC96-0106-1
113: AC81-0580-6, AC91-0121-23
114: AC93-0150-114, AC93-0186-52
115: AC93-0186-103, AC84-0712-15
116: AC95-0154-119, AC95-0154-186
117: AC80-0512-3
118: AC94-0034-18, AC92-0323-144
119: AC88-0620-24, AC87-0031-1, AC89-0557-5
120: AC86-8010-6, AC95-0088-1
121: AC91-0121-22
122: A74-4515, AC93-0237-120, AC94-0480-29
123: AC86-8016-7, AC82-0875-155, AC82-0875-149
124: AC95-0203-51, AC95-0203-39
125: A70-2957, AC95-0203-17, 12-ft logo
126: AC79-0126-1, AC95-0272
127: A79-0054-4, AC93-0132, AC99-0101-3
128: AC-31030, AC96-5007-12
129: A-25685-3, A-25685-14
130: AC80-0570-3, AC76-1060, AC75-1002, A-26796, A-27773
131: AC87-0594-15, AC-42628-1
132: AC90-0448-22, ACD97-0151-1 (*Tilt Rotor Research Aircraft*, by Attila Hejja)
133: AC85-0186-16, AC84-0473-100
134: AC78-0579-1
135: AC71-3921
136: AC80-0455-6, AC95-0403-1
137: AC89-0246-8
138: AC87-0091-5, AC80-0613-3
139: AC71-2429
140: AC80-0864-4, AC82-0253-14
141: AC82-0778-1, AC92-0323-128
142: AC95-0275, AC93-0593-3
143: AC85-0569-8, AC79-0403-2
144: AC73-5172-6, ACD93-0456-10, AC96-0327-3
145: AC82-0121-5, AC95-0029-156

146: AC75-0205, 80-HC-625 (*Tile Team*, by Morton Kunstler)
147: AC75-1161-4, AC73-2590
148: AC91-0397-1
149: 89-HC-628 (*Belly of the Bird*, by Deborah Deschner), AC92-0127
150: Shuttle, AC71-4076, Gap Fillers, FRCI-12, AFRSI
151: A-31494, AC76-0564
152: AC77-0475-10.1 (Art by Don Davis), A90-3006, AC91-3008
153: AC76-1011-2-33, A-34401
154: 89-HC-271 (*Viking 2 Passes Over Mars*, by Lonny Schiff)
155: AC78-9245 (*Pioneer Venus Multiprobe*, by Paul Hudson)
156: AC78-9247 (*Pioneer Galileo Separation*, by Paul Hudson)
157: AC85-0354-6, AC83-0831-2
158: AC81-0483-1, 92-HC-139 (*Galileo Test*, by Henk Pander)
159: AC82-0516-19
159: AC88-0552-8
160: AC77-0068-27, AC-42233-6
161: AC95-0294-3
162: AC89-0114-580, A-70-5719-2-8
163: AC87-0912-6, AC72-2246
164: AC90-0361-8, AC80-0006-2.1
165: AC80-0767.2, Leonid Expedition
166: AC83-0768-10, AC72-5317
167: AC83-0768-2.1, AC83-0768-2.1

168: AC88-0473, AC94-0048-4
169: ACD96-0069-14, AC88-0327-3
170: AC81-7046, AC91-0099-1, AC91-0257 (Art by Don Davis)

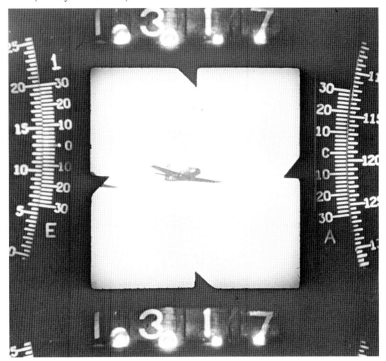

171: AC79-7061, 96-HC-427 (*Eyes on the Environment*, by Bryn Narnard), ACD95-0024.1
172: AC84-0378-2, AC86-0345-9, AC88-0477-3
173: A76-1366-1, A94-0009-1
174: AC85-0145-12, AC87-0395-4
175: AC85-0742-30, AC80-0382-15
176: AC77-1181-9, A-41652
177: AC92-0552-282, AC92-0401-2
178: A93-0511-7, AC93-0230-25

179: G92-47-231-30, AC97-0106-1
180: AC98-0244-590, AC96-0254-1
181: AC72-4372.1, A-24321
182: A-28284, AC88-0327-2
183: AC81-0712
184: NAS Time Line (Art by James Donald)
185: AC84-0373-15
186: AC87-0093-3
187: AC93-0146-3, AC88-0514-40
188: F-18, AC86-0541-17, AC89-0187-2
189: AC91-0365-14
190: AC91-0016-8
191: AC95-0396-2
192: AC88-0149-2.1
193: AC95-0395-1
194: AC95-0400-1
195: AC95-0401-1
196: AC86-0583-10, AC86-0583-12
197: Nanocarbon, Nanogear, AC86-0583-3, AC95-0393-1
198: AC93-0559-1, AC92-0007-5
199: AC95-0423-6
200: AC92-0352-2, AC95-0054-7
201: AC93-0051-3, AC87-0662-38, AC94-0261-7
202: AC94-0261-8, AC89-0437-20, ACD96-0205-1
203: 93-HC-7 (*Virtual Reality*, by Andreas Nottebohn)
204: AC86-0189-73
205: AC99-0222-4
206: A89-0786-1
207: AC99-0222-5
208: AC97-0219, A89-0787-13

Chapter 4

210: Human Centered Computing (Copyright PhotoDisc)
211: Archaea
212: AC94-0125-2, AC97-0316-22
214: AC95-0280-6
215: AC93-0121-1, AC91-0121-1, AC91-0121-15
216: 81-HC-16, AC92-0408-8
217: A91-0102-21, AC96-0204-1
218: AC94-0183-15
219: AC99-0222-6
220: AC93-0627
221: AC96-0386-22, AC76-1091-10
222: AC94-0281-1, AC98-0160-1, AC94-0273-41
223: ACD97-0042-15, AC94-0206
224: ACD99-0137-3, AC99-0130-17
227: AC95-0220
228: AC99-0141
229: AC96-0392-8, ACD99-0180-2
230: ACD97-0063-2, ACD98-0200-2
231: AC96-0400-3, AC97-0249-7, NREN Patch (Art by Thomas Esposito)
232: ACD97-0367.1, AC93-0610
233: AC95-0054-112, AC97-0289-1
234: ACD98-0078-4

235: AC94-0195

236: AC96-0058-1, AC91-0005-18

237: AC96-0347-2, AC99-0095-1.1

238: AC97-0295-12

239: AC97-0295-1, AC97-0295-13

240: Astrobiology Art (Art by Cheryse Triano), AC96-0345-9, AC86-0614-28, ACD97-0336, AC96-0345-3

241: ACD96-0290-2

242: AC96-0110-1, AC94-0398-1

243: ACD99-0103-3, AC92-0401-3

244: AC99-0142-1, AC97-0195-22

245: AC99-0177-8, AC96-0252-16

247: AC96-0196-13

248: AC97-0388-1.2, ACD97-0047-2.1

249: ACD98-0027-3, ACD97-0047-6

250: ACD97-0047-4

251: AC96-0151, EC97-44294-2

252: AC94-0199-18

253: ACD98-0042-1

254: AC93-0284-7

255: AC99-0174

256: CASC

257: ACD98-0165-1

258: AC97-0316-247, AC97-0316-215

Appendix

260: AC99-0222-8

266: AAL-6013

267: EL-1996-00158

Index

A-26B, 4, 19
Abbott, Ira, 55, 80
Ackerman, Thomas, 155, 169
AD-1 oblique wing, 108, **109**, 113
Advanced X-ray Astrophysics Facility, 212
Aerojet-General, 87
Aerospace operations systems, 236-240
AETB (alumina enhanced thermal barrier) material, 149
AFRSI (advanced, flexible, reusable surface insulation), 148, **150**
AH-1G, 141, 142, 246
Aiken, Edwin, 143, 246
Airborne science, 158-163
Airbus A300, 112
Air Force Manned Orbiting Laboratory, 183
Air Force Satellite Control Facility (Onizuka Air Force Station), 219
Air traffic control, 236-240
Alger, George, 159
Allamandola, Louis, 171
Allen, H. Julian, 2, 13, 22, **24**, **28**, 28-32, 30, 32, 46, 48, **50**, 51, 52, 57, 60, 66, 80, **85**, 86, 102, **103**, 105
  as Ames director, 84-97
  and blunt-body concept, 30, 59, 60-62
Ames, Joseph Sweetman, vii, 11, 255, **260**, 260-267, **267**
  Ames Aeronautical Laboratory named for, 261
  and Johns Hopkins University, 262-264
  as NACA founding member, 264-267
  and National Research Council, 264
Ames Aeronautical Laboratory, 1, 5
  founding of, 5-12
  and hypersonic studies, 46-48
  site selection criteria for, 7-8
  and supersonics, 32-40
  and transonic studies, 44-46
  and Unitary plan wind tunnel facility, 40-44
Ames Aerospace Encounter, **105**, 107, 220
Ames Basic Research Council, 99
Ames Research Center
  and aerodynamics studies, 68-74
  Aerospace Encounter, **105**, 107, 220
  aerospace operations systems, 236-240
  airborne science studies, 158-163
  in aircraft safety research, 116-120
  aircraft transfers from, 55-56, 226
  in air traffic control research, 236-240
  in arc jet development, **62**, 64-65, **65**
  in astrobiology, 240-243
  astrochemistry research at, 168-169
  in astronaut suit design, 73-74
  and atmospheric reentry studies, 59-68
  and boundary layer control, 37
  CFD at, 191-196
  in computational chemistry, 196-198
  computer applications, 181-204
  consolidation of Dryden Flight Research Center, 107-108
  Cosmos/Bion missions, 174-178
  Crew Vehicle Systems Research Facility, 106
  deicing research at, 14-16
  Equal Opportunity Trophy awarded to, 218
  exobiology research at, 167-168
  "faster, better, cheaper" projects, 246-254
  flight research at, 109-128

fly-by-wire research at, 110-112
and Galileo Jupiter probe, 156-158
in heat-protective materials development, 146-150
and hypersonic aerodynamics, 60-62
hypervelocity free-flight facility, 62-64
information technology at, 228, 229-235
infrared astronomy at, 163-167
intelligent systems research, 198-204
jet aircraft research, 34-40
as lead Center in astrobiology, 242
as lead Center in Earth observation aircraft, 160-161
as lead Center in gravitational biology and ecology, 243
life sciences at, 57-58, 74-80
in lifting body design, 81-82
and lunar sample analyses, 78-80
Multi-Cultural Leadership Council, 217-218
NASA security check at, 213-214
in parallel supercomputing, 185-187
from past to future, 254-259
and Pioneer program, 87-97
and Pioneer Venus program, 153-156
planetary science at, 150-172
relations with NASA Headquarters, 51-59, 213-216
in rotary-wing aircraft development, 141-145
SETI program, 172-174
and simulator development, 70-72
and space program management, 80-83
and space shuttle development, 145-150
and supersonic research, 32-40
in supersonic wing design, 34-38
telepresence development, 200-204
TQM at, 216-217
and university collaborations, 103-104
in VTOL aircraft design and development, 128-136
wind tunnel upgrades, 120-128

in X-36 development, 250-252
ZBR, 224-228
Ames Strategy and Tactics Committee, 99
AME-2, 229
Anderson, Dale, 234
Anderson, Seth, **31**, 37
Andrews, Richard, **234**
Apollo program, **55**, 66, 69
navigation, **72**, 72-73
magnetometers, 79-80
space capsule, **66**, 76
Arc jet heaters, **62**, 64-65, **65**, 147
Argonne National Laboratory, 197
Arizona, University of, 165
Armstrong, Sam, 229
Army Aeroflightdynamics Directorate, 141, 246
Army Air Corps, 6, 9
Army Aviation Research and Development Laboratory, 132
Arnold, Henry H., 9, 10, 11, 12, 267
Arnold, James, 187, 196-197
Arnold Engineering Development Center, 115
ARPA (Advanced Research Projects Agency), 104, 189
Artificial intelligence. *See* Intelligent systems
Asteroid belt, 94-95
Astrobiology, Ames as lead Center in, 1, 220, 240-243
Astrochemistry, 168-169
Astronaut suits, 73-74
Astronomical Society of the Pacific, 254
Atkinson, David, 157
Atlas Centaur, 94
Atomic Energy Commission, 53
Automation Sciences Research Facility (ASRF), 204
AutoNav, 234
AV-8B, **131**, 131
Aviation Safety Reporting System (ASRS), 117-118

Aviation System Capacity Program, Ames as lead Center in, 236

B-2, 114
B-17, 15
B-24, 15
B-32, 18
B-58, 37
B-70, 69, 105
Bader, Michael, 164
Bailey, F. Ron, 189, 205
Bailey, Rodney, 251
Ballhaus, William F., Jr., 114, **196**, **205**, 227
    as Ames Director, 204-207
Barrillearx, James, **163**
Base Realignment and Closure Commission (BRAC), 214
Bauschlicher, Charles, 197
Bay Area Economic Forum, 227
Beech Aircraft, 194
Bell Aircraft Corporation, 129, 133, 135
Belsley, Steve, 30
Benn, Ross, **24**
Berry, William E., 180, 227, 229, 255, **255**, 256, **257**
Betts, Edward, **24**
Billingham, John, **57**, 172, 173, **173**
Billings, Charles, 117
Binder, Allen, 248
Bingham, Nancy, 126, 225, 227, **257**
Bioletti, Carlton, **24**, 25, 33, 81
Biology, gravitational. *See* Gravitational biology
Biosatellite program, 81
Bisplinghoff, Ray, 84
Black, David, 168
Blake, David, 167
Blocker, Alan, **24**

Blumberg, Baruch S., **243**, 243
Blunt-body reentry design, 30, **58**, 59-62
Blunt trailing edges, 37-38
Boeing Aircraft, 112, 133, 138, 139, 194
Boeing 707, 36
Boeing 747, 252, **253**
Boese, Robert, 154, 157
Boundary-layer control, 37
Bousman, William, 144
Boyd, John W., 36, **37**, 85, 107, 229, **257**
Braig, Eugene, **24**
Braig, Raymond, **24**
Braxton, Lewis S.G., III, **257**
Bregman, Jesse, 166
Brennwald, Louis, 107
Briggs, Geoffrey, 233
Bright, Loren, 69, 85, 184-185
Brooks, Elmer T., 217, 218
Browning, Rowland, **24**
Bryson, Steve, **202**
BTD-1, 19
Buck, Andre, **24**
Bulifant, George, **24**
Bull, Jeff, 149-150
Bunch, Theodore, 167
Burchard, Karl, **24**
Burgess, Virginia, **24**
Burroughs Corporation, 184, 186
Burrous, Clifford, **86**
Bush, George, 208, 212
Butow, Steve, **165**

C-8 turboprop, 138
C-130, 150, **159**, 159, 223, 226
C-141, **164**, 165, 226. *See also* KAO
Cabrol, Natalie, **208**

California Air and Space Center, 258
California Air National Guard, 222
California Institute of Technology, 10, 37
Campbell, Thomas, 206
Canning, Thomas, **63**, 63-64
Cassini missions, 170
CDC (Control Data Corporation), 186, 187
Center for Bioinformatics, 231-232
Center for Mars Exploration (CMEX), 233-234
Center of Excellence, 220, 225
Center for Star Formation Studies, 168
Center for Turbulence Research, 192
Center TRACON Automatic System. *See* CTAS
Central Computer Facility, 183
Centrifuge, 127
CFD (computational fluid dynamics), 191-196
    in F-111 design, 193
    in HiMAT design, 114
    in space shuttle design, 193
Chambers, Alan, **102**
Chambers, Lawrence, 177
Chang, Sherwood, 167
Chapman, Dean, 30, 37-38, 65, 66-68, **67**, 102, 147, 184-185, 197
    and atmospheric reentry studies, 60, 62
Charters, Alex, 48
Chartz (Smith), Marcie, **181**, **182**, 182
Cheng, Rei, **231**, **232**
CH-46, 135
CH-47, 141, 142-143, **143**
Chin, Benny, 157
Civil Aeronautics Authority (CAA), 12
Clark, James, 201
Clarke, Frank, **24**
Clark-Y airfoil, 36
Clearwater, Yvonne, 203

Clementine spacecraft, 247
Cleveland Clinic Foundation, 232
Clinton, William J., 214
Clousing, Lawrence, **19**, 20, **27**, 30
CMEX (Center for Mars Exploration), 233-234
Cockpit resource management, 118-119, 236-238
Cochran, John, 139
CoE-IT (Center of Excellence for Information Technology), 228, 229-235
Coleman, Jana M., 216, 220, **257**
Colin, Larry, 157
Collier Trophy, 1, 15, 265
Colombano, Silvano, 199
Compton, Dale L., **206**, **207**, 207-209, 213, 214, 215, 216, **217**, 218-219, 227
Compton Gamma Ray Observatory, 212
Computational chemistry, 189, 196-198
Computational fluid dynamics. *See* CFD
Computer, 5
    Alliant FX/8, 187
    CDC Cyber 205, 187
    CDC 7600, 186
    Connection Machine, 187
    Convex C-1, 187
    Cray 1S, 187
    Cray-2, 188
    Cray X-MP/22, 187
    Cray X-MP/48, 187
    Cray Y-MP, 188
    Cray Y-MP C90, 188
    Electrodata Datatron 205, 182
    Honeywell 800, 183
    IBM 360/50, 183
    IBM 1800, 183
    IBM 7040, 183
    IBM 7094, 183

Illiac IV, 104, 181, 182, 184-186, 192
Intel Hypercube, 187
Computer applications, 181-204
Computer History Museum, 258
Concorde, 69
Condon, Estelle, 163
Conical camber, **33**, **36**, **37**, 37
Connolly, Jack, 179
Consolidated Supercomputing Management Office (CoSMO), 230
Convair 990, 159, 164. *See also* Galileo II
Cook, Arthur, 7, 11
Cook, Woodrow, 37
Cooper, Chris, **227**
Cooper, David, 198, 228
Cooper, George, 20, **26**, **31**, 37
Cooper-Harper Handling Qualities Rating Scale, 20
Cordova, France, 241
Corliss, Lloyd, 142
Cornell Aeronautical Laboratory, 20
Cosmos/Bion missions, 174-178, 244
cancellation of, 178
Cowings, Patricia, **175**
Crane, Robert, 80
Cray, Seymour, 188
Creer, Brent, **72**
Crew Vehicle Systems Research Facility. *See* CVSRF
Crows Landing (Naval Auxiliary Landing Field), 221-222
Cruikshank, Dale, 170
CTAS (Center TRACON Automatic System), 237-238
Culbertson, Phil, **202**
Cultural Climate and Practices Plan, 220
Cuzzi, Jeffrey, 170
CVSRF (Crew Vehicle Systems Research Facility), 106, 119, 236-237
Cyclops project, 172

Dalton, Bonnie, **151**, 179
DARPA (Defense Advanced Research Projects Agency), 140, 184, 189
Daugherty, Thomas, 248
Davies, M. Helen, **24**, 30
DC-8, **113**, 226
Dean, William, 219, 223
DECnet, 190-191
DEEC (digital electronic engine control), 111
Deep Space Network, 173
Deep Space 1 spacecraft, 234
Defense Advanced Research Projects Agency. *See* DARPA
DeFrance, Smith J., 5-6, **10**, **11**, 11, 13, 14, 17, **22**, **24**, 30, 32, 48, 49, 51, 52, 56, 58, 77, 80, 81, 82, 86, **182**, 182, 206, 255
as Ames director, 23-25
Deicing research, **14**, 14-16
Delaney, John, Jr., **24**
Delaney, Noel, **24**
Denning, Peter, 187
Dennis, David, 45
Department of Defense, 136, 247
Department of Transportation, 117
Des Marais, David, 167
Deutsche Akademie der Luftfartforschung, 266
DeVincenzi, Donald, 168
DeYoung, John, 37
DFBW (digital fly-by-wire) research, 110-112
Diaz, Al, 241
Digital Equipment Corporation (DEC), 190
Digital fly-by-wire technology. *See* DFBW
Dimeff, John, 48, 76, 150
Dive flaps, 17-18
Drake, Frank, 173
Drinkwater, Fred, **57**, 130

Dryden, Hugh, 42, 51, 55, 263
Dryden Flight Research Center, 82, 105, 107-108, **108**, 114, 115, 145, 225, 226
   in fly-by-wire development, 110-112
Duct rumble, 18
Dugan, Daniel, 142
Duller, Charles, 160
Dunlap, Herbert, **24**
Durand, Willam F., 13, 25, **266, 267**, 267
Dusterberry, John, 13, 71, **102**
Dyal, Palmer, 79
Dynamic stability testing, 38-39

Earth Observation System satellite (EOS), 160
Earth Resources Technology Satellite (ERTS), 160
Edwards Air Force Base, 55, 107
Eggers, Alfred, 30, 45, 60, 80-81, **83**, 84, 105
80 by 120 foot wind tunnel, **114, 115**, 121-128
Eisenhower, Dwight D., 49, 51, 54
Ellis, Stanley, 175
Ellis, Stephen, 202
Equal Opportunity Trophy, 218
Erhart, Ronald, 134
Erickson, Albert, 17
Erickson, Edwin, 252
ER-2, **112**, 160, 161-162, **162, 163**, 226
Erzberger, Heinz, **236**, 237
Eshoo, Anna, 226-227
Eshow, Michelle, 143
Espinosa, Paul, **132**
European Space Agency, 244
Evans, Bradford, 56
Exobiology, 77-80, 167-168

F-8, 110, 112
F-15, 111, 112, 116
F-16, 111
F-86, 21, 37, 55
F-100, 21, **26**
F-101, 39
F-102, 37
F-104, 21, 57, **83**
F-111, 113
F/A-18, 116
FAA (Federal Aviation Administration), 104, 111, 120, 136, 138, 173, 236-240
Fakespace Corporation, 232
Falarski, Michael, 221
Falsetti, Christine, **231**, 231
Farmer, Jack, 167
"Faster, better, cheaper," 1, 153, 212, 230, 246-254
Feldman, William, 248
Fettman, Martin, 180
Few, David 133
F4D-1, 28
Fibrous refractory composite insulation (FRCI-12), 149
Fisher, Scott, 201
Flap, rotating cylinder, 137
Flap-suction study, 37
Flightpath management, 111-112
Flight research at Ames, 20-21, 26-28, 68-70, 109-128
   instrumentation, 114-116
   vehicle, remotely augmented, 114-115
Flight Simulator for Advanced Aircraft (FSAA), **76**, 116-117, 246
FLITE (Flying Laboratory for Integrated Test and Evaluation), 116-117, 246
Ford, Kenneth, 230, **257**
40 by 80 foot wind tunnel, **17, 18, 23**, 37, **118**
   upgrading of, 121-122
   in VTOL testing, 130
Foster, John D., 142

Foster, John V., 83, 88
Foster, Mayo, 24
FRCI-12 (fibrous refractory composite insulation), 149
Freeman, Arthur, 24
Frick, Charles, 24, 33, 39
Friedland, Peter, 199
F6F-3, 20, 21
Future Flight Central facility (FFC), 237, 239

GALCIT (Guggenheim Aeronautical Laboratory of the California Institute of Technology), 8-10
Galileo I, 160, 165
Galileo II airborne laboratory, 159-160, 164, 165
Galileo Jupiter probe, 156-158, 241
GASP (General Aviation Synthesis Program), 194
Gault, Donald, 68, 150
Gawdiak, Yuri, 229
General Aviation Synthesis Program (GASP), 194
Gerbo, Carl, 24
Gerdes, Ronald, 142
German aeronautical research, 6-7, 254, 266
Givens, John, 157, 245
Glazer, Jack, 103-104
Glennan, T. Keith, 51, 53
Glenn Research Center. *See* Lewis Research Center
GLOBE (global backscatter experiment), 162
Goddard Space Flight Center, 26, 53, 105, 107, 154, 160, 226, 247
Goett, Harry, J., 19, 22, 24, 26-28, 27, 30, 70, 80
   as director of Goddard Space Flight Center, 53
Goldin, Daniel, 212, 221, 226, 229, 241-242, 243, 243, 253, 259
   as NASA administrator, 212-219
Goldstein, Howard, 147, 149, 178
Goody, Richard, 153
Goodwin, Glen, 65, 65

Gordon, Helen, 45
Gore, Albert, Jr., 214
Graeber, Curtis, 236
GRAPES project, 162
Gravitational biology, 178-181, 243-246
Greene, Mark, 24
Griffin, John, 231
Grindeland, Richard, 175
Guastaferro, Gus, 108, 157, 248
Guerrero, Michael, 230

Haberle, Robert, 169
Hall, Charles F., 33, 33, 36, 37, 81, 83, 87, 91
   in Pioneer programs, 88-93, 154
Hall, Warren, 140
Halley's comet, 155-156
Handling qualities research, 20-21, 131
Hansen, Robert J., 229, 230, 257
Hardy, Gordon, 227
Harke, Stanton, 239
Harper, Charles W., 37, 53, 132, 193
Harper, Lynn, 180, 258
Harper, Robert, 20
Harper, William, 30
Harrier (AV-8B), 131, 131
Hart, James, 190
Hartsfield Atlanta International Airport, 238
Hartman, Edwin, 25
Harvey, Charles, 24
Haymaker, Webb E., 57
Heaslet, Max, 30, 35, 45, 46
Heinle, Donovan, 28
Heitmeyer, John, 36
Hertzog, Alvin, 24
Hewlett-Packard Company, 172, 248
HIDEC (highly integrated digital electronic control), 111

High-lift devices, 37
High-Resolution Microwave Survey (HRMS), 174
HiMAT (highly maneuverable aircraft technology test bed), 108, 114
Hine, Butler, 234
Hines, John, **180**, 181
Hinkley, Robert H., 12
Hinshaw, Carl, 10
Hirschbaum, Howard, **24**
HL-10 lifting body, 82
Hogan, Robert, **206**
Hollenbach, David, 171
Holloway III, James L., 135
Holton, Emily, 175
Holtzclaw, Ralph, 88
Hood, Donald, **24**
Hood, Manley, **24**, 30
Hoover, Herbert, 265
Horne, Clinton, **122**
Houston, John, **24**
Howe, John, 61
Hubbard, G. Scott, 227, 233, 241, 242, **248**, 248, 249
Hughes, Robert, **24**
Hughes Aircraft, 157
Human factors research, 75-76, 117, 203
Human Performance Research Laboratory (HPRL), 203
Hunsaker, Jerome, 10, 41
Hunten, Donald, 154
Hunter, Lisa, **222**
Huntress, Wesley, 241
Huntsberger, Ralph, 42
Hypergravity test facility, 74, 127
Hypersonics, 46-48, 60-62, 105, 149
Hypervelocity Free-Flight Facility (HFF), 62-64, **63**, 152

IBM computers, 183
Illiac IV supercomputer, 104, 181, 182, 184-186
  in sonic boom studies, 192
Illinois, University of, 184
Independent Verification and Validation Facility, 235
Information Technology, Center of Excellence for, (CoE-IT), 228, 229-235
Infrared astronomy, 163-167, 207
Inlets, submerged, **39**, 39-40
Internet, 230-232
Ikeya-Seki comet, 164
Ilyin, Eugene A., 177
Innis, Robert, 139
Institute of Advanced Computing, 195-196
Institute for Biomedical Problems, Moscow 175
Institute for Space Research, 241
Intelligent systems, 1, 198-204
International Astronomical Union, 168
International Geophysical Year, 53
IRAS (Infrared Astronomy Satellite), 166-167, 207
ISO 9001 certification, 255-256

Jacklin, Stephen, 246
Jaffe, Richard, 197
Jahnke, Linda, 167
James, Donald, 258
JASON project, 220
Jedlicka, James, 157
Jemison, Mae, 180
Jet Propulsion Laboratory. *See* JPL
Johns Hopkins University, 261, 262-264
Johns Manville, 148
Johnson, Kelly, 17
Johnson, Lyndon B., 67, 206
Johnson Space Center, 75, 147, 225, 226

Jones, Alun, 30, 33
Jones, Lloyd, **102**
Jones, Robert T., 33-34, **34**, 35-36, 45, **102**, 102, **103**, 109
   and oblique wing, 112-123
JPL (Jet Propulsion Laboratory), 93, 97, 158, 166, 173 174, 209, 226, 249, 253
   and MESUR mission, 233
JUH-60A, 143-144
Jupiter probe. *See* Galileo Jupiter probe

Kalman filter, 72-73, 142
Karman, Theodore Von, 9
Kasting, James, 155, 169
Katzen, Elliot, 36
Kaufmann, William, 20, 21
Keller, Richard, 240
Kelley, P. X., 135
Kelly, James, **24**
Kelly, Mark, 121
Kennedy, John F., 54, 55
Kerwin, William, 48
Keyser-Threde GmbH, 254
Kidwell, George, 227
Kiris, Cetin, 195
Kittel, Peter, 252
Klein, Harold P., 77, **151**, 153
Klineberg, John, 105
Kliss, Mark, 181
Knutson, Martin, 160, 226, 227
Kolodziej, Paul, 149-150
Konoplic, Alex, 249
Koop, Mike, **165**
Kourtides, Demetrius, 120
Kufeld, Robert, 144
Kuiper Airborne Observatory (KAO), 165-166, 253

Kuiper, Gerald P., 165
Kutler, Paul, **196**, 205
Kwak, Dochan, 195

Landsat, 162
Langhoff, Stephanie, 197
Langley Aeronautical Laboratory, 5-6, 7, 17, 42, 44, 265
Langley Research Center, 17, 55, 105, 140, 226
Larson, Howard, 146, 147
Laser speckle velocimetry, 120
Lasinski, Thomas, 190
Lead Center for Astrobiology, 242
Lead Center for Gravitational Biology and Ecology, 243
Learjet, **111**, 159, 165, 226
Lear Siegler Corporation, 111
Lebacqz, Victor, 142
Lehmann, John, 135
Leiser, Daniel, 149, **178**
Leon, Henry, 175
Leon, Mark, 234
Lepetich, Joseph, 88
Levit, Creon, 190
Lewis, George W., 265, **267**
Lewis Research Center, 42, 55, 105
Lichten, Robert, 129
Liewar, Walter, **28**
Life sciences research, 57-58, 74-80
Lifting bodies, **59**, 81-82, **83**, **84**
Line-oriented flight training (LOFT), 118
Lindbergh, Charles, **10**, 12
Lissauer, Jack, 170
Lockheed Missiles and Space Company, 49, 161, 248, 249
Lockheed 12A, **14**
Lomax, Harvard, 35, 45, 181, 182, 183, 191
Los Alamos National Laboratory, 248

Lovelace, A. M., 107
LRSI (low-temperature reusable surface insulation), 148
Lunar and Planetary Laboratory, University of Arizona, 165
Lunar Prospector mission, 220, 247-250
Lunar Research Institute, 248
Lum, Henry, **197**, 198, 204, 228

McAvoy, William, **19**, 30
MacCormack, Robert, 191
MacCreight, Craig, 252
McDonald, Henry, vii, **228**, 235, **243**, 254, 256, **257**, 259
    as Ames director, 228-229
McDonnell Douglas, 111, 129, 194, 250
MacElroy, Robert, 181
McGreevy, Michael, 200
McKay, Christopher, 167, **223**, 234
McNair Intermediate School, **217**
Macomber, Thomas, **24**
Magellen Venus, 209
Magnetometers, **79**, 79-80, 91
Mah, Robert, 176, **230**
Mandel, Adrian, 175
Man Technology, 254
Mark, Hans, 2, 86, 93, **100**, **102**, **103**, **104**, 106, 133, 135, 160, 168, 172, 184, 189, 196, 197, 225
    as Ames director, 99, 100-104
    in Pioneer program, 88
Marlaire, Michael, **218**, 227, 256
Marshall Space Flight Center, 226
Mars Environmental Survey (MESUR), 233
Mars Pathfinder, 233
Martin, James, 139
Massa, Manfred, **24**
Massachusetts Institute of Technology, 199

MD-900 helicopter, **221**
Mead, Merrill, **54**, 85
Meeker, Gabriel, **242**
Merrill, Robert, 142
Mersman, William, **181**, 182
Meyer, George, 142
Meyers, Glenn, 232
Microgravity, research into effects of, 178-181, 244-246
Migotsky, Eugene, 36
Millikan, Clark, 9
Millikan, Robert, 9, 10
Minden, Lysle, **24**
Mineta, Norman, 214, 225
Miya, Eugene, 190
Modern Rotor Aerodynamic Limits Survey (MRALS), 144
Moffett, William A., 8, 267
Moffett Field, 7, **8**, 8, 56
    reconfiguration of, 214-215, 220-222
    as Moffett Federal Airfield, 222, 223
Mohri, Momuro, 180
Moralez, Ernst, 144
Morrison, David, **173**, 241, 253, **257**
Morse, David, 258
Mort, Kenneth, 122
Mossman, Emmet, 39, 40
M2-F2 lifting body, 82, **83**, 105
Mulenburg, Gerald, 127
Munechika, Ken K., **219**, 219-224, 228
Munk, Max, 35, 265
Murphy, James, 252

NACA (National Advisory Committee for Aeronautics), 1, 5
    aeronautical research facilities survey committee, 12
    Ames, Joseph S., as founding member of, 264-267

Future Research Facilities Committee, 7
in GALCIT tunnel proposal, 10
in hypersonic research, 46-48, 49
in supersonic research, 32-40
in transonic research, 44-46
transition into NASA, 51-54
in Unitary Plan wind tunnel proposal, 40-42
NAI (NASA Astrobiology Institute), 220, 242-243
NAS (Numerical Aerospace Simulation facility), 106, 187-189, 205, 230, 232
NASA (National Aeronautics and Space Administration)
   Astrobiology Institute (NAI), 220, 242-243
   Center for Bioinformatics, 231-232
   Deep Space Network, 88, 94, 173
   Discovery Program, 247
   Exceptional Service Medal, 106
   founding of, 51
   Goldin, Daniel, as administrator of, 212-219
   Independent Verification and Validation Facility, 235
   Millennium program, 234
   Mission Analysis Division, 105-106
   Mission to Planet Earth, 162
   Office of Advanced Research Programs, 55
   Office of Advanced Research and Technology (OART), 58, 84, 120-121
   Office of Aeronautics and Space Technology (OAST), 99, 205
   Office of Space Science and Applications (OSSA), 83
   and Pioneer program, 86-97
   and proposed Moon mission, 54-55
   relations with Ames Research Center, 51-59, 213-216, 224-228
   security review of Ames, 213-214
   SETI program, 173-174
   Western Aeronautical Test Range, 107
   in X-36 tailless research aircraft development, 250-252

   Zero Base Review, 224-228
National Academy of Sciences, 172-173
National Advisory Committee for Aeronautics. *See* NACA
National Aerospace Plane, 193
National Air and Space Museum, 97
National Full-Scale Aerodynamics Complex. *See* NFAC
National Hispanic University, 220
National Performance Review, 224
National Research and Education Network (NREN), 231
National Research Council, 77, 169, 264
National Security Agency, 173
Navajo Nation, 232
Naval Air Reserve at Moffett Federal Airfield, 222
Naval Observatory, 262
Naval Research Laboratory, 53
NC-130, **31**
Neel, Carr, 150
Nettle, Mildred, **24**
Neurolab mission, 244-245
NFAC (National Full Scale Aerodynamics Complex), **114**, 116, 121, 123, 124, 145
Nickle, Ferril, **12**, 13, **24**
Nielsen, Jack, 34, **103**
Nimbus meteorological satellite, 80
Nissen, James, **19**, 20
Nixon, Richard M., 104
Nottebohm, Andreas, **202**
NREN (National Research and Education Network), 231
Numerical Aerospace Simulation facility. *See* NAS

O-47A, **9**
Oblique wing, **102**, **109**, 112-113
O'Brien, Bernard, 92
O'Brient, Thomas, **24**
O'Hara, Dolores, **178**, 178

Oliver, Bernard, **172**, 172, 173
1 by 3 foot supersonic wind tunnel, 32
Onizuka Air Force Station, 219, 222
Orbital Sciences, Inc., 194
Orbiting Astronomical Observatory (OAO), 80
Ortega, Toni, **222**
Outdoor aerodynamic research facility (OARF), **116**
OV-10, tunnel testing of, 137, **139**
Oyama, Jiro, 76, **176**,
Oyama, Vance, 57, 79, **153**, 153, 154

P-3 Orion, 165, 214, 221, 222, 223
P-38, 17-18
P-47, 18
P-51 Mustang, 18, 19, **33**
P-61 Black Widow, **19**, **43**
P-80 Shooting Star, **29**, 34
PAET (planetary atmosphere experiments test), 152
Pappas, Constantine, 61
Parker, John, 119, 120
Parsons, John F., **10**, **12**, 13, **22**, 22, **24**, 26, **30**, 42, 49, 57, 75, 80, 82, 85
    as Ames associate director, 25-26
Pennsylvania State University, 195
Peterson, Victor L. 69, 207, 218-219
Peterson, Walter, **24**
Philip C. Crosby, Inc., 216
Philpott, Delbert, 175
Pioneer International Quiet Sun Year probe (PIQSY), 81
Pioneers
    and Apollo mission, 87
    Emmy award for, 95
    6 and 9, 86-88
    10 and 11, 88-97
Pioneer Venus, 153-156, **154**
    and Halley's comet, 155-156

    multiprobe bus, **155**
Pipkin, Roselyn, **24**
Pirtle, Melvin, 195
Planetary atmospheres, studies of, 151-152
Planetary exploration, telepresence in, 233
Planetary science, 150-172
Pollack, James, 155, 157, **169**, 169, 170
Ponnamperuma, Cyril, 77, 77-78, 79
Poole, Manie, **24**
Powered-Lift Flight Research Facility, 139. *See also* QSRA.
Presley, Leroy L., 194
Prizler, Paul, **24**
Program management, **54**, 80-83
Proxmire, William, 173
Pueschel, Rudolf, 162-163
Pulliam, Thomas, 192

QSRA (Quiet Short-Haul Research Aircraft), 137-139, **138**
Quaide, William, **68**, 79

Radioisotope thermoelectric generator (RTG), 91
Ragent, Boris, 154, 157
Rai, Man Mohan, 195
Randt, Clark, 57
RASCAL (Rotorcraft Aircrew Systems Concepts Airborne Laboratory), 143-144, 246
Rasky, Daniel, 149-150
Rasmussen, Daryl, **232**
Raymond, Arthur, 41
Raytheon E-Systems, 253
RCG (reaction-cured glass), 147-148
Reagan, Ronald, 107, 108, 175
Reentry aerodynamics, 66-68
Reentry heating, 60-62, 64-65
    and ablative materials, 64

arc jets in simulating, 64-65
    thermal protection, 146-150
Reese, David, 151
Reeves Electronic Analog Computer (REAC), 182
Reinath, Mike, **119**
Remotely augmented flight-test vehicle, 114-115
Research Institute for Advanced Computer Science (RIACS), 187
Reynard, William, 117, 118
Reynolds number, defined, 21
Reynolds, Ray, 151, 168, 169, 170
Riddle, Dennis, 139
Roberts, Leonard, **103**, 109
Robinson, Russell, **11**, 13, 25, 52, 80, **208**
Rodert, Lewis, **15, 24**
    Collier Trophy awarded to, 15
    in deicing research, 14-16
Rodrigues, Annette, 221
Rogallo, Robert, 186
Rogers, Stuart, 195
Rogers Dry Lake, 19, 55, 107
Roland, Alex, 265
Rolls, Stewart, **28**
Roosevelt, Franklin D., 11, 261, 265
Rosen, Robert, 216, **257**
Ross, Muriel, **231, 234**
    in bioinformatics, 231-232
    in Neurolab mission, 245
ROSS (Reconstruction of serial sections) software, 232
Rossow, Vernon, 150
Rotor, free-tip, 141-142
Rotorcraft Aircrew Systems Concepts Airborne Laboratory. *See* RASCAL
Rotor System Research Aircraft. *See* RSRA
Rotary wing aircraft, 141-145, 246
Rowland, Henry, 263

RSRA (Rotor Systems Research Aircraft), 137, **140**, 140-141
Rubesin, Morris, 61

Sabreliner, 114
Sadoff, Melvin, **57**
Sagan, Carl, 93-94, **173**, 173
St. John, Marie, **24**, 30
St. John, Robert, **74**
Salama, Farid, 166
Salinas Medical Center, 232
Sanders, Douglas, **132**
Santa Clara University, 103
Savage, Tad, 180
Schiff, Lewis, 116
Schmidt, Gregory, 179
Schmidt, Stanley, 72-73
Schnitker, Edward, **24**
Schroeder, Jeffery, 143
Seiff, Alvin, 47, 63-64, 151, 152, 153, 154, 157
Sensor 2000! program, 181
Serrano, Rick, **255**
SETI (search for extraterrestrial intelligence), 172-174, 218
SETI Institute, 173, 174, 254
7 by 10 foot wind tunnel, 18, 132
    in engine inlet studies, 39-40
    in stall studies, 37
Sharp, Edward, 24
Shiner, Robert, 237
Shockey, Gerald, 142
Short, Barbara, 47
SH-3G, **110**, 141
Sikorsky, 129, 140
Silicon Graphics, Inc., 190, 201, 232, 239, 243
Silva, Richard, **86**

Sims, Mike, 202
Simulator, **135**
    Apollo navigation, **72**
    Boeing 747-400 cab, 236
    five-degree-of-freedom, **69**, **70**, 71
    FSAA, **76**, 116-117
    HICONTA, **51**
    six-degree-of-freedom, **71**, 71
Simulators
    early development of, **69**, 70-72
    in flight training, 118
SIRTF (Space Infrared Telescope Facility), 209, 252
6 by 6 foot supersonic wind tunnel, 33, 39
16 foot high-speed wind tunnel, **13**, **16**, 16-17, 34
Skin friction measurements, 38
Slotnick, Daniel, 184
SLS-1. *See* Spacelab Life Sciences mission
SMA (surface movement advisor), 238
Smith, Donald, 38
Smith, Gerald, 73
Snowden, Cedi, **201**
SOFIA (Stratospheric Observatory for Infrared Astronomy), 252-254
Somerich, E., **31**
Sommer, Simon, 47
Sonett, Charles P., 57, 150
Souza, Kenneth, 177, 179, **258**
Space Act agreements, 103-104
Space Camp California, 220
Space capsule, Gemini, 76
Space Infrared Telescope Facility. *See* SIRTF
Spacelab Life Sciences missions, 179, 180
Space Shuttle, 57, 119, 145-150
    Challenger, 157
    heat shields, 146-150
    orbiters, 105, 145, 149

Space Station Biological Research Facility (SSBRF), **243**, 245-246
Space Station, International, 208, 224, **241**, 244, 245-246
    Mir program, 244
Spacesuits, 73, 73-74, **258**
Space Technology Laboratories, 81
Sperans, Joel, 157
Sperry Rand, 133
Spreiter, John, 35, 44, 45, 150
SSBRF (Space Station Biological Research Facility), **243**, 245-246
Stack, John, 44
Stadler, Jackson, 27
Stanford University, 192, 207, 257
    in early Ames staffing, 13
    Medical Center, 232
Statler, Irving, 141
Sterling Software, 201, 232, 253-254
Stevens, Kenneth, 187
Stevens, Victor, 145
Stine, Howard, 157
Stoker, Carol, **198**, 234
STOL (short takeoff and landing) aircraft, 136-137
Stollar, Lee, 125
STOVL (short takeoff and vertical landing) aircraft, 137
Strategic Defense Initiative Office, 212
Stratospheric Observatory for Infrared Astronomy, 252-254. *See also* SOFIA
Sun Microsystems, 190, 191
Sun, Sid, **177**
Supersonic free flight tunnel, 44, 45, 47
Supersonic transport (SST), **68**, 69, 105
Supersonic wing design, 34-38
Supersonic wind tunnels, 41-44
Surface movement advisor (SMA), 238

SV-5D lifting body, 82
Swain, Robert, 124
Swept wing, 35
Syvertson, Clarence A., 4, 45, 84-85, **106**, 121, 204, 207, **208**
    as Ames director, 105-109
    and SETI program, 173
Szalai, Kenneth, 226

TAAT (tip aerodynamics and acoustics test), 142
Tarter, Jill, **172**, 174
Tavasti (Lt. Col.), **31**
TCP/IP data transfer protocol, 190-191
Teacher Resource Center, 220
Tektites, 66-68, **67**
Teller, Edward, 184
Terminal radar approach control (TRACON), 237-238
Thermal protection materials for atmospheric reentry, 61, 146-150
3.5 foot hypersonic wind tunnel, **60**, 105, 145
Tielens, Xander, 171
Tiles, thermal protection, 147-148
Tilt rotor aircraft, 132-136
    Model I-G, 128-129
    V-22, 135-136
    XV-15, **101**, 132-134
Tobak, Murray, 39
Tomko, David, 179
Toon, Brian, 163, 169
Toscano, William, **175**
Toughened uni-piece fibrous insulation. *See* TUFI
Towers, John, 12
TQM (Total Quality Management), 216, 217, 255
TRACON (terminal radar approach control), **236**, 237-238
Tran, Huy, 150

Transcendental Company, 128
Transonics, 44-46
    area rule, 36
    drag rise, 33
Trippensee, Gary, 111
Truly, Richard H., 206, 208, 212, 215
Truman, Harry S., **15**, 41
TRW Inc., 92, 212
TUFI (toughened uni-piece fibrous insulation), 149
Turner, William, 20
Tverskaya, Galina, 177
12 foot pressure wind tunnel, **20, 21**, 21-22
    upgrade of, **125**, 126
20-g centrifuge, 179

U-2 in basic research, 160
UH-1H, 141, 142, **143**
UH-60, **142**, 144, 246
Unitary Plan wind tunnel facility, 26, **40**, 40-44, 127-128, 145
United Airlines, 240, 254
Universities Space Research Association (USRA), 187, 253
University of California, 254
    at Berkeley, 168
    at Davis, 162
    Radiation Laboratory, 76
    at Santa Cruz, 168, 257
U. S. Air Force, 160
USRA (Universities Space Research Association), 187, 253
USS *Kitty Hawk*, QSRA testing aboard, 139
USS *Macon*, **6**
USS *Tripoli*, 135
U.S. War Department, 40

Valero, Francisco, 162
Van Dyke, Milton, 32, 35, 45
Vanguard project, 53
Variable-stability aircraft, development of, 21, 115
VAX computer, 190
VENERA spacecraft, 155
Venus probe. *See* Pioneer Venus
Vernikos, Joan, 175, **178**, 178
Victory, John, 11, 265, **267**
Videll, Lesslie, **24**
Vincenti, Walter, 13, **24**, 44, 207
Virtual visual environment display (VIVED), 200
Visitor Center, 220
VMS (vertical motion simulator) upgrade, **126**, 127
Vojvodich, Nicholas, **157**
Vostok spacecraft, 176
Vought (LTV), 131
VPL Research, 200
V/STOL (vertical/short takeoff and landing) aircraft, **130**, 131-132
V/STOLAND flight guidance system, 142
VTOL (vertical takeoff and landing) aircraft, 122, 128 136
   computer codes for, 195
   V-22 tilt rotor, 135-137
Vykukal, Hubert, 73, 73, **76**, **201**, 203

Walker, Wilson, **24**
Wallops Flight Center, 107, 152
Washington, University of, 120
Watson, Edie, **104**
Webb, James E., 54
Webbon, Bruce, 181
Welch, Robert, 179
Wenzel, Elizabeth, 202
Wernicke, Ken, 133

Western Aeronautical Test Range, 107
Westover, Oscar, 6
Wettergren, David, **208**
White, John, 73
Whiting, Ellis, 197
Willey, Margaret, **24**
Wilson, Charles, 83
Wilson, Clyde, **24**
Wilson, Woodrow, 264
Wind tunnel
   High Reynolds number channel, **117**
   10 by 14 inch, 105
   1 by 3 foot supersonic, 32, **35**
   3.5 foot hypersonic, 105, 145
   6 by 6 foot supersonic, 33
   7 by 10 foot, 18, 37, 39-40, 132
   12 foot pressure, **20**, **21**, 21-22, 38
      upgrade of, **125**, 126
   16 foot high speed, **13**, **16**, 16-17, 18, 34
   40 by 80 foot, **17**, **18**, 18-19, **23**, 37, **56**, **118**, 121
   80 by 120 foot, **104**, 119, 121-128
      accident at, 123-124
   Unitary plan, 26, 40-44, 145
      upgrade of, 127-128
Wind tunnels
   data collection, 114-116
   testing technologies, 114-116
   upgrading of, 120-128
Wisnieski, Richard, 225
Wolfe, John, **86**, 88, **96**, 150, **173**
Wood, Donald, 29, 30
Woodrum, Clifton A., 11, 266
World War II wind tunnel research, 16-22
Wray, Alan, 186
Wright, Orville, **267**

X-2, 49
X-3, 130
X-5B, **131**
X-14, 21
X-14B, **130**
X-15, **49**, 70-71
X-36, 250-252, **251**
XC-142, 131
XFR-1, 19
XF7F-1, 19
XP-51, **13**
XSB2D-1, 18
XV-1, 129
XV-2, 129
XV-3, **129**
XV-5A, 131
XV-6A, 131
XV-15, **101**, **132**, 132-134, **133**
Xenakis, George, 142

Yee, Helen, 192
YF-93, **39**
YO-3A, 109-110, **110**
Young, John, 57, 145-146, 157, 169
YP-80, **29**
YROE-1, **128**

Zell, Peter, **122**
Zero Base Review (ZBR), 224-228, 229
Zornetzer, Steven F., 229, **257**
Zuk, John, 136
Zweben, Monte, 199

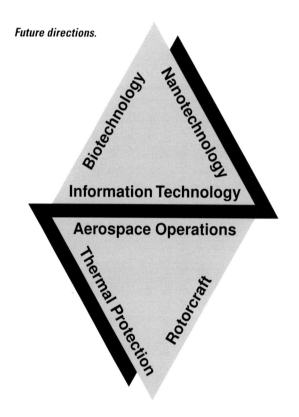

Future directions.

☆USGPO 584-146 2000